INDIGENOUS STUDIES AND
ENGAGED ANTHROPOLOGY

Advancing the rising field of engaged or participatory anthropology that is emerging at the same time as increased opposition from Indigenous peoples to research, this book offers critical reflections on research approaches to-date. The engaged approach seeks to change the researcher-researched relationship fundamentally, to make methods more appropriate and beneficial to communities by involving them as participants in the entire process from choice of research topic onwards. The aim is not only to change power relationships, but also engage with non-academic audiences.

The advancement of such an egalitarian and inclusive approach to research can provoke strong opposition. Some argue that it threatens academic rigour and worry about the undermining of disciplinary authority. Others point to the difficulties of establishing an appropriately non-ethnocentric moral stance and navigating the complex problems communities face. Drawing on the experiences of Indigenous scholars, anthropologists and development professionals acquainted with a range of cultures, this book furthers our understanding of pressing issues such as interpretation, transmission and ownership of Indigenous knowledge, and appropriate ways to represent and communicate it. All the contributors recognise the plurality of knowledge and incorporate perspectives that derive, at least in part, from other ways of being in the world.

The book includes powerful insights from different parts of the world that demonstrate the challenges, nuances, ethics, practices, potential and power of engaged approaches. Relationships between researcher and researched communities are skilfully negotiated through individual chapters and the dialogue that was created through the symposium to bring these pieces together. Such work is never comfortable, and dialogue can be testing and difficult but it is worth the effort on all sides as it uncovers understandings that could not come to be in any other way. Thoroughly recommended for those beginning their journey of engagement and for those who are well along the way.
                    Linda Tuhiwai Smith, The University of Waikato, New Zealand

Sillitoe has (again) hit upon a collection that wonderfully meets an urgent contemporary need for anthropologists. His wide range of contributors address a bevy of pertinent issues and one can only hope that the collaboration they advocate persists beyond the present moment of interest.
                    Joy Hendry, Oxford Brookes University, UK

This important and timely book provides answers to the pressing question that now confronts young anthropologists setting off to do fieldwork with indigenous peoples: 'what is the benefit of this to us?' By bringing together contributions from indigenous scholars and anthropologists this book enriches both fields and provides a framework for dialogue directed to realizing the emancipatory possibilities of research that is done with people rather than about them.'
                    Paul Oldham, United Nations University, Japan

Indigenous Studies and Engaged Anthropology

The Collaborative Moment

Edited by

PAUL SILLITOE
Durham University, UK

Routledge
Taylor & Francis Group

LONDON AND NEW YORK

First published 2015 by Ashgate Publishing

2 Park Square, Milton Park, Abingdon, Oxfordshire OX14 4RN
52 Vanderbilt Avenue, New York, NY 10017

Routledge is an imprint of the Taylor & Francis Group, an informa business

First issued in paperback 2020

British Library Cataloguing in Publication Data
A catalogue record for this book is available from the British Library

The Library of Congress has cataloged the printed edition as follows:
Indigenous studies and engaged anthropology : the collaborative moment / edited by Paul Sillitoe.
    pages cm
  Includes bibliographical references and index.
  ISBN 978-1-4094-4541-8 (hardback : alk. paper) 1.   Indigenous peoples--Research--Methodology. 2. Applied anthropology. 3. Ethnology--Methodology. I. Sillitoe, Paul, 1949- editor of compilation.
  GN380.I537 2014
  305.8'00723--dc23

                                                                          2014018357

ISBN 978-1-4094-4541-8 (hbk)
ISBN 978-0-367-66899-0 (pbk)

In support of the diverse voices of indigenous peoples whose fight for their integrity and for their rights against neocolonial practices challenge and expand our understanding of the world, and of ourselves.

Contents

List of Figures

Notes on Contributors

Domenica Gisella Calabrò studied for a PhD as a member of the research programme 'Anthropologies, Institutions, Representations' in the Faculty of Education at the Università degli Studi di Messina, Sicily, Italy. She investigated the indigenization of rugby in New Zealand and its role in the process of Māori identity definition. She was hosted by 'Te Kawa a Māui', the School of Māori Studies at Victoria University of Wellington, during her fieldwork.

Emma Cervone is a cultural anthropologist and Associate Director of Latin American Studies at John Hopkins University. She has worked on the indigenous movement in Ecuador and has an active research agenda in collaboration with indigenous organizations on issues of racism, gender and indigenous justice in Ecuador. She is the author of the book *Long Live Atahualpa, Indigenous Politics, Justice and Democracy in the Northern Andes*.

George J. Sefa Dei is currently Professor of Social Justice Education, Ontario Institute for Studies in Education at the University of Toronto. His teaching and research interests are in the areas of anti-racism, minority schooling, international development, anti-colonial thought and indigenous knowledge systems. He is a traditional chief in Ghana, being the Adumakwaahene of the town of Asokore, and his stool name is Nana Sefa Tweneboah I.

J.P. Linstroth is an Affiliate Research Professor with the Department of Anthropology at Florida Atlantic University. He obtained his D.Phil. in Social and Cultural Anthropology from the University of Oxford. Also, he was a recipient of a J. William Fulbright Foreign Scholar Grant (2008–2009) to study urban Amerindians in Manaus, Brazil and to be a Visiting Professor with the Department of Anthropology at the Universidade Federal do Amazonas (UFAM).

Raymond Nichol is Co-ordinator International and Senior Lecturer, Social Science Education, in the Faculty of Education, La Trobe University, Victoria, Australia. He is a Fellow of the Australian Anthropological Society. In teacher education he specializes in the fields of humanities, citizenship and Indigenous studies and education, particularly in Australia and Melanesia. He is the author of *Growing Up Indigenous: Developing Effective Pedagogy for Education and Development* (Sense Publishers, 2010).

Jayantha Perera was Principal Safeguards Specialist of the South Asia Department at the Asian Development Bank, Manila, the Philippines. He is currently Visiting Professor of Development Studies at the Kelaniya University, Sri Lanka. His key areas of interest are involuntary resettlement, indigenous peoples, environmental law and practice, international law, and agrarian change in South Asia. He is a fellow of the Royal Anthropological Institute, London.

Robyn Sandri is a descendant of the Kooma-Gungarri peoples of South West Queensland. She completed a PhD in 2012, gathering the stories of four generations of story tellers and their experiences with colonial schools. Her work is focused on the role of story to the maintenance of a culture across time and how the loss of story disrupts the trans-generational knowings and identity of colonized Aboriginal cultural groups. Robyn lectures in Indigenous Studies at Australian Catholic University in Sydney.

Priscilla Settee is Associate Professor in the Department of Native Studies at the University of Saskatchewan and a member of Cumberland House Cree First Nations from northern Saskatchewan. In 2012 she received the University of Saskatchewan Provost's Award for Teaching Excellence in Aboriginal Education, and in 2013 she was awarded the Queen Elizabeth Diamond Jubilee Award for contributions to Canada. She is author of *Pimatisiwin, Global Indigenous Knowledge Systems* (John Charlton Publishing Ltd, 2013) and *The Strength of Women, Ahkameyimowak* (Cocteau Books, 2011).

Rachel Shah is a PhD student in the Anthropology department, Durham University. Her current research is enquiring into, and contrasting, traditional informal learning and formal school education in the highlands of Papua, Indonesia. Her research interests include an interrogation of issues of international development, education and social justice in the context of indigenous knowledge and culture.

Paul Sillitoe is Professor of Anthropology, University of Durham. His interests include sustainable development and social change, human ecology and ethno-science, livelihood and technology, with research experience in the Pacific, South Asia and Gulf regions. His current research focuses on natural resources management, technology, and development, and he is engaged in work on conservation in Qatar.

Dimitrios Theodossopoulos is Reader in Social Anthropology at the University of Kent. His research addresses the topics of exoticization, ethnic stereotypes, indigeneity, authenticity, resistance and protest, the economic crisis, and the politics of cultural representation in Panama and Greece. He is author of *Troubles with Turtles* (Berghahn, 2003), editor of *When Greeks Think about Turks* (Routledge, 2006), and co-editor of *United in Discontent* (Berghahn, 2010) and *Great Expectations* (Berghahn, 2011).

Tran Tran is Research Fellow based at the Australian Institute of Aboriginal and Torres Strait Islander Studies (AIATSIS). Tran completed her PhD at the UNESCO Centre for Water Law Policy and Science, University of Dundee. Her thesis was about the influence of knowledge structures on water management in the Canning Basin, Western Australia. She has worked in the Native Title Research Unit at AIATSIS focussing on recognition and security of Indigenous land and water rights and supporting Indigenous governance structures. Tran is currently working on the design and implementation of collaborative management arrangements over Indigenous lands.

Preface

Many people around the world increasingly oppose research in their communities. The hostility often relates to the perception that research, undertaken archetypically by anthropology, goes hand-in-hand with wider political-economic domination and inequality, fostering accusations of neocolonialism. It is an aspect of mounting criticism of prevailing academic practice and calls for a radical change in the conduct of research, that it should feature more egalitarian and inclusive approaches. This more inclusive way of doing research goes back to the 1960s when methodologies seeking to change the relationship between researcher and researched started tentatively to appear. These aim to revise the research process so that it is more responsive to the needs and priorities of communities, involving them as participants, from choice of research topic and key questions, to the publication of results. This way of doing research, variously called action or participatory research, increasingly goes under the label of engaged anthropology. It aims not only to change power relationships within research projects, but also improve engagement with the wider public and interdisciplinary collaboration.

Another response to local opposition to research comes from within communities themselves, which may seek to undertake their own research. It occurs increasingly under the label of Indigenous studies. It predictably covers a broad spectrum of interests with native scholars seeking to advance understanding of knowledge in their locales, which globally vary widely. These Indigenous studies are distinctive in that they aim to tackle problems identified by communities and to undertake research that will benefit them. The methods are also distinctive, in often genuinely seeking to involve all those concerned as equal partners.

While both engaged anthropology and Indigenous studies seek to undertake participatory research that is appropriate and assists communities, their perspectives differ strikingly in some regards. The form of knowledge they focus on often varies and their ideas about its role in contemporary society. Anthropologists largely consider the content and structure of local perspectives according to Western intellectual principles of logical coherence. They focus, for instance, in so-called indigenous knowledge enquiries, primarily on peoples' understanding of their environments and how they gain their livelihoods and treat illness, in their endeavours to contribute to development efforts to extend the benefits of industrial technology, underpinned by agricultural and medical science, to poor communities, by including local expertise and concerns more meaningfully and equitably in the process.

Some Indigenous scholars oppose such trends, arguing that these destroy the cultural and environmental heritage of their communities. The shift implies, at best, hybridization of local knowledge, or at worst, its destruction, and simultaneously the integration of communities more extensively into the global economy. The relevance of certain aspects of traditional Indigenous knowledge and practices to contemporary life is an issue here. These native researchers more often emphasize the moral and spiritual dimensions of knowledge, which they may draw on in struggles to achieve human rights and self-determination, environmental equity, access to land and resources, and ultimately cultural survival.

In order to establish a truly engaged anthropology, it is necessary to address these disconnections between approaches and methods, which challenge anthropology's fundamental claim to represent the lifeways and worldviews of others. Indeed interaction between Indigenous studies and engaged anthropology has the potential dramatically to alter the practice of socio-cultural anthropology. The translation of non-Western ways of being in the world into terms understandable by a Western audience is a core part of anthropology. The extent to which these translations distort their subject matter has long been a topic of debate, sharpened in recent decades by the postmodern critique. The meaningful inclusion of the translations and critiques of Indigenous scholars into this debate can only enrich and deepen it: in what ways do their perspectives add to, and modify, our understanding of indigenous ontologies and epistemologies?

In some ways, the methodologies of Indigenous studies match those of engaged anthropology. While Indigenous scholars are drawing on experience as the 'researched' in formulating their egalitarian methodologies, and anthropologists are building upon experience as 'researchers' in developing participatory methods, both feature inclusive approaches to enquiries that make a difference locally. It is disconcerting, then, that instead of mutually profiting from interaction and discussion of their approaches and perspectives, they are developing largely in parallel. Although integration promises to benefit both, up to now anthropologists have largely failed to include Indigenous academics in their debates, while Indigenous scholars are often suspicious of anthropologists. This volume seeks to contribute to the breaking down of the barriers.

The concerns of Indigenous studies inform, to a greater or lesser extent, all of the contributions to the book. They recognize the plurality of knowledge traditions and seek open exchange, incorporating perspectives that have derived, at least in part, from other ways of knowing. Themes addressed include how to tackle the dominance of anthropology, and Western social science generally, in accounts of Indigenous peoples. Another is the implication for anthropology of equalizing research arrangements, including the discipline's place within the academy. The interrogation of ideas of Indigenity, their representation and anthropology's role are further topics. The political issues of representation and ownership of knowledge emerge as key themes.

Some Indigenous authors voice concerns that foreign researchers extract, modify and co-opt knowledge from their communities with no return. These contentions prompt questions about how much better indigenous scholars' representations are – which presumably differ to some extent too from local community views, as informed (polluted) by the expectations of formal non-Indigenous education – and what are appropriate ways of representing and communicating Indigenous knowledge and concerns that may vary from region to region. What are the implications for local knowledge, often tacit and experiential, of eliciting and presenting it according to academic canons, even if these differ from those of western universities? What can the perspectives and experiences of indigenous scholars teach us about the way we represent and pass on knowledge: should we start thinking about teaching anthropology in new ways?

While the aims of engaged anthropology seem positive, it has nonetheless provoked strong opposition from within the academy, some arguing that it threatens analytical rigour and may compromise scholarship. If the research agenda is set through a participatory process it may dilute disciplinary expertise by preventing researchers from focussing on topics they specialize in. Other political concerns, particularly with respect to action research, include the difficulty of navigating the complex problems that many, often marginalized communities face, and the danger of reinforcing inequalities by aligning, albeit inadvertently, with more powerful factions. These problems in turn relate to the challenge of establishing an appropriately non-ethnocentric moral stance, which further relates to issues of representation, domination and ownership of knowledge.

It is to such pressing and exciting issues that this book addresses itself, bringing together a unique combination of views to further discussion between Indigenous studies and engaged anthropology, and strive for a more inclusive and egalitarian anthropology. The chapters present diverse and sometimes contradictory perspectives on, and across these themes, drawing on the experiences of Indigenous scholars, social anthropologists and development professionals acquainted with a range of cultures in various regions of the world.

It is a pleasure, as always, to close this preface with an acknowledgement of thanks. All contributors acknowledge a debt of gratitude to Serena Heckler for initiating the discussion that led to this book. It started when she and I had tenure, as junior and senior partners, of a Nuffield Foundation New Career Development Fellowship (NCF/32406). We thank the Nuffield Foundation for its generous support, and the British Academy, too (Grant CSG-52196). Finally, I thank Jackie Sillitoe for running her laser-like copyediting pen over the manuscript to ensure that all the contributions followed the publisher's house style consistently.

Note

The sharp-eyed will notice that the word indigenous is sometimes spelled indigenous with a low case 'i' and other times as Indigenous with a capital I. This is a political issue for some contributors, who consider Indigenous equivalent to adjectives such as European or American in distinguishing certain populations. (Hendry and Fitznor (2012: xiii) note the same apparent inconsistent spelling of a range of words, asking the reader to understand that currently 'there would seem to be little global agreement' over their usage.)

References

Hendry, J. and Fitznor, L. (eds) 2012. Introduction, in *Anthropologists, Indigenous Scholars and the Research Endeavour: Seeking Bridges Towards Mutual Respect*, edited by J. Hendry and L. Fitznor. Abingdon: Routledge, 1–18.

Chapter 1

The Dialogue between Indigenous Studies and Engaged Anthropology: Some First Impressions

Paul Sillitoe

When I was engaged in some teaching at the University of Papua New Guinea (PNG), I had an awkward experience. A colleague became increasingly irate during a conversation after a seminar. It was over a book that had recently appeared that was proving – bafflingly, to my colleague – influential among anthropologists. He told me that he could not recognize much about his life and ideas as a Papua New Guinean in the book; in fact frankly he could not understand it at all really. It sowed a seed, albeit at the time, as often happens, I did not realize that it had been planted in my mind. It was further stimulated to germinate when I heard Roger Keesing describe, and subsequently I read, how when the Kwaio people, who live on Malaita in the Solomon Islands, agreed to him living with them, they did so thinking that he would support 'their efforts to codify their customary law and ancestrally imposed taboos so as to demand their recognition by the state' (1992: vii). An expectation that he declared was a 'curse' because he could not write such a 'customary lawbook … from the critical standpoint of an anthropologist' (1992: 13), and he never attempted to do so, leaving the project to 'the Kwaio themselves' (1992: 13).

It seems odd that an anthropologist should declare that he could not engage with what was of interest and concern to the people he lives with, because it is not relevant from his research perspective (Morauta (1979) made a similar observation some time ago). Or that a discipline that purports to further understanding of other cultural ways of being in the world should produce work in which the subjects themselves cannot recognize their behaviour or ideas (see Owusu 1978: 312; Nakata 1998; Lassiter 2005: 121). The nagging worries prompted by these two experiences and others are, for me, behind this volume, which seeks to engage with such issues, and help us to see beyond, and work towards resolving, such contradictions.

The Society Concept and Domination

The notion of society – which was central to the book that so infuriated my PNG colleague – illustrates the nature of such contradictions. It is an idea that has long presented us with problems. The people I know in the Papua New Guinea

Highlands not only apparently have no such concept (and certainly no word that comes anywhere near it, to my knowledge) but also behave in ways that make it difficult to identify such a meta social collectivity (for instance, all those who speak the same language – who we might consider as constituting a society – have no collective name for themselves and differ in social behavioural particulars from place to place (Sillitoe 1979, 2010)). Regardless of the apparently imposed, 'second intention' status, of the term society, social scientists assume that all human beings must live in a society – that we find such everywhere – that persons interact as collectivities and observe certain agreed norms to guide their behaviour – or else their disciplines have no meaning (Tuhiwai Smith 1999: 47–50). Indeed it is this concept that underpins their claim to some intellectual authority and ability to further our knowledge and understanding of human interaction and behaviour.

Where does this put people such as my highland New Guinea friends who do not seemingly have such an idea? In a word, it puts them at a disadvantage when faced with social scientists, who claim to use this concept and a barrage of related ones to understand their behaviour. But we find anthropologists in an odd position too – as I have pointed out previously (Sillitoe 2007: 157–8) – of appearing to know more about the people they study than they know about themselves, when the object of study is their ideas and behaviour – which is of them. Some social scientists are astonishingly, indeed arrogantly confident in the supremacy of their knowledge, as evident in the following comment criticizing a paper that I submitted to a journal, where I remarked on the propriety of anthropologists assuming to represent a superior understanding:

> I disagree strongly with the statement: "It is no longer tenable – if it ever was – for us to represent the lifeways, beliefs, etc., of those we study. Most populations are able to represent themselves". The literature is full of examples of literate populations not understanding their sociocultural system. The locals are not aware of the problems that the anthropologists address and the uses of their work; there is an "unconsciousness" in sociocultural systems and the locals do not understand these; the locals do not fully understand the interplay of their own meanings nor see the interrelations between the domains of their sociocultural system.

It is reminiscent of Malinowski's (1922: 11–12) comments about 'depicting the constitution' of Trobriand society where the problem was that the 'natives obey the forces and commands of the tribal code, but they do not comprehend them' and consequently 'it would be futile to attempt questioning a native in abstract, sociological terms', which we might take as early authorization for the imposition of the concept of 'society' and associated social science 'theory', or an admission of the inevitability of Westerners introducing such foreign notions to make sense of what they observe, hear and experience. It is time to move beyond the participant-observation approach that his work heralded towards participatory-reflection (Sillitoe 2012), collaboratively formulating new approaches that allow the 'natives' to express their understandings.

What does this suggest about the discipline of anthropology? It seems to confirm, as some indigenous scholars in particular argue, that it serves, wittingly or unwittingly, as an agent of Western domination. It has scarcely done so by acting as the 'handmaiden of colonialism', as some ill-informed critics suggest (Sillitoe 2006: 7) – anthropologists having contributed relatively little to colonial administrations in their domination of subject peoples, being largely considered whacky, irrelevant outsiders (much as they are in some development circles today). But the discipline has acted as a notable force for Western domination of others by forcing understanding of their lifeways to fit Western concepts, to serve the intellectual concerns and agendas of Western authorities (see, for example, Banerjee's and Linstead's 2004 critique of the 'indigenous land ethic' concept). The Canning Basin case study discussed by Tran Tran in this volume illustrates the consequences for Aboriginal people in Australia, of how government administered legal recognition of native title to land and water resources dispossesses and distorts native views and rights.

But to play devil's advocate, which is my intention in this introduction to stir up interest, it is arguable, taking a postmodern line, that such an outcome was inevitable. How else might we expect these Western intellectual intruders to make sense of communities elsewhere, already socialized as they were into an understanding of the world, except according to the concepts and knowledge they had inherited from their culture? This is an inescapable bugaboo that stalks us, for no matter how much we struggle to see that our work is as true as we can make it to those we seek to understand, it is a vain hope from a postmodern viewpoint (Flaherty et al. 2002). So maybe it is inevitable that we focus on concerns that may be irrelevant to those we live with and produce work in which they cannot recognize their behaviour or ideas? If so, all the more reason for a dialogue with, or better still collaboration with indigenous scholars. For this is surely one way to break out of the postmodern impasse.

Distinguishing 'Terms of Intention'

On what grounds has Western academic discourse assumed authority for its concepts and understanding (Tuhiwai Smith 1999: 59–65)? Unlike technological inventions, from which the domination of Western culture largely stems – the diffusion of which to the presumed betterment of humankind informs international development[1] – Western philosophical knowledge is not demonstrably superior. When we take a closer look at the notion of society and other such ideas, we see the questionability of their assumed authority. The concept of 'society', a 'term of second intention' in logic (as formulated by the Scholastics of the Middle Ages), is one for which we have no empirical sense derived evidence, unlike a

---

1   I am speaking historically here and leaving aside the increasingly evident environmental problems that industrialization brings.

'term of first intention' such as 'human individual', which refers to someone/ thing that we can see, touch, hear, smell etc. (see Russell 1946: 463–4; Jacob 2003; *Encyclopædia Britannica* 2009). While 'terms of first intention' concern particular unique examples of phenomena – such as anyone reading this book – to which we can refer in our discussions and compare our ideas whatever our socio-cultural background (assuming that all humans have the same senses); 'terms of second intention' concern universalizing abstractions – such as a group of socially interacting human beings – afford no concrete referent as abstract ideas (I cannot sit my idea of English society in a chair to investigate it further, unlike my brother).

The origin of such abstract concepts is a philosophical conundrum going back at least to Plato's metaphysical 'Theory of Forms', which asserts that for ideas represented by general terms such as justice, beauty etc., there exist somewhere beyond our ken timeless, ideal, mind-independent, supernatural-power-[God]-created entities. But it appears that the 'Theory of Forms' ideal types differ between intellectual traditions, these culturally informed general terms being subject to wide cultural-linguistic variation, which as second order intellectual constructions present us with abstract categories that are particularly difficult to access, translate and correlate cross-culturally. And we certainly cannot rank them, one superior to another, except on subjective criteria according to your culturally informed values. What is encoded and thought about as a category in one language – say a collection of interacting human beings comprising a society descended from a troupe of hairy apes – may be encoded quite differently in another – say a clan that according to origin myths came from a flock of sulphur-crested cockatoos.

The so-called theories of social science are consequently socio-culturally and historically situated constructions, more so than natural science theory, which focuses experimentally on 'terms of first intention' phenomena and uses rigorously defined and universally agreed 'terms of second intention' to interpret results. The predictable consistency of the things comprising many scientific 'terms of second intention' facilitates such theorizing – at least on planet Earth – one molecule of CO_2 is constituted exactly the same and behaves identically to any other – at least in the macro-world we inhabit. It is the control, and scope to interfere in the natural world, which this predictability allows, that underpins aforementioned scientific technology that shores up Western domination. Human behaviour, as we know, does not show such consistency, which compounds the problems faced by social science because not only are its 'terms of second intention' culturally and historically contingent but its 'terms of first intention' are individually highly variable too! The social sciences are only now facing up to the implications of individual behavioural variation with the move from a structural to a processual focus, challenging assumptions about homogenous normative codes guiding social interaction or members of the same social group sharing the same concepts, values etc. You have only to ask a sample of persons met in the street what they understand by the word 'society' to see this, for you will receive a range of differing answers – that is, those of us who have the concept of 'society' cannot even agree what it is (not even social scientists!).

Tackling Dominance: Ideology versus Theory

The difference between the 'terms of intention' that the natural and social sciences investigate has prompted me to argue that it is more accurate to talk about social science ideology than theory (Sillitoe and Bicker 2004). The implications of admitting that social science deals in ideology not theory are considerable. It undermines dominance of Western social science. It indicates that there are no objective grounds for declaring one view better than another; it is a subjective matter informed by culturally shaped values that can vary widely (Stanfield 1985). And it reveals the part anthropology has played in furthering domination by intellectual means, contributing to the proliferation of social science theories – that are subject to rapid fashion changes that seem uncannily to track ideological fashions in Western society – which the discipline consequently imposes somewhat dubiously elsewhere, employing them as if they have the stamp of some unquestionable universal academic authority.

Whatever one's view of the inevitability or otherwise of this process of intellectual domination following European exploration and intrusion into other parts of the world, it is increasingly acknowledged that it is time we sought to give an equal hearing to other voices and views (Chambers 1997; Lamphere 2003: 153). We need to embark on a journey together.[2] Do people entertain notions equivalent to society, for example, and how do they express these, and how may these be more effectively used in interacting with authorities that interfere in their lives? If they do not entertain such ideas, it is not our place, I think, to try and imagine what the idea of society might be in its apparent absence – even if justified as a provisional concept (because as a foreign idea to start with it can never be replaced with a final more appropriate one). It is again to impose the concept in another guise, when we should allow people to express their understandings of human social interaction. What other ways are there to conceive alternative 'terms of second intention' for collective social behaviour? They may, for instance, involve unfamiliar notions of reciprocity, certainly in the New Guinea highlands.

Rather than defending theoretical-cum-ideological turfs, anthropologists should increase efforts to make joint cause with indigenous scholars to secure their ideological views an equal airing (Cervone 2007), which should be easy for those in a discipline predicated on the premise that we have much to learn from other ways of being in the world. But instead, it is difficult, for it implies a dramatic change in current academic arrangements, with a genuine democratizing of power structures – an equal 'interweaving' of indigenous and anthropological perspectives, after symmetrical Tongan *felavai* patterns (Ka'ili 2012) – whereas Western scholars are no different to others who hold power in being reluctant to relinquish it, even though the intellectual rewards may be great. But events are increasingly forcing such change, particularly the increasing irritation, even hostility, shown by communities to outsider researchers, notably in the Americas but also increasingly elsewhere

---

2   Hendry and Fitznor, and colleagues (2012) use the metaphor of bridge-building.

such as Australasia (Rigney 1999, 2006; Greene 2004, McCarty et al. 2005: 1), which are fed-up with foreign nosey-parkers who contribute nothing in return. As an Australian website for indigenous studies comments, 'we have been researched to death and there is nothing on the ground to show for it ... why don't they leave us alone?' (Umilliko Indigenous Higher Education Research Centre n.d.). In some instances, people see researchers perpetuating exploitative neo-colonial relations, as agents of wider political-economic forces widely glossed as globalization, which are resulting in considerable unwelcome upheaval in some regions. The sentiments are common in Latin America, as the infamous El Dorado affair discussed by Emma Cervone graphically illustrates, and the chapter by John Linstroth shows further, recounting the struggles of Amerindians living in the Brazilian city of Manaus and how the expectation from the start of his research was that he would reciprocate their help by collaborating with them in their political battle with the urban authorities.

Indigenous Collaboration

There have been some moves towards greater inclusivity. One example is the indigenous knowledge in development initiative that builds on participatory approaches that many anthropologists support, which seeks to include the expertise, viewpoints and aspirations of local communities meaningfully and equitably in the development process (Warren, Slikkerveer and Brokensha 1995; Antweiler 1998; Grenier 1998; Sillitoe 1998; Ellen and Harris 2000; Emery 2000). While the indigenous knowledge approach focuses largely on understanding and management of local environments, how people secure their livelihoods and treat sickness, it seeks to encompass socio-cultural context, including consideration of political issues and history, which often feature unequal power relations that perpetuate poverty. It has been dogged by the manipulation of participatory approaches in development contexts, as those involved in projects seek to reach the 'management milestones' set down by accounting fixated political paymasters (Cooke and Kothari 2001; Mosse 2005). The indigenous scholarship approach holds out the hope of seeing currently abused participatory research methods applied properly.

Another example is the way in which some museum ethnographers have sought in the last decade or so to collaborate with curator colleagues and local communities in regions from whence their collections originate, in arranging displays and documenting objects (Herle with Bani 1998; Simpson 2001; Hendry 2002; Peers and Brown 2003; Stanley 2007). Consequently, ethnographic museums have moved from turgid backwater to the anthropological mainstream, introducing fresh ideas and leading the way in participatory research. It may be that dealing with 'term of first intention' things – namely objects – gives them an edge; as it is easier to collaborate with things – such as masks – to focus discussions on, rather than abstract ideas – such as society. There is something nicely ironic in this, as museum ethnography's focus on things contributed – following early Victorian evolutionary interest – to it being considered an uninspiring and largely ignored aspect of anthropology, until recently (Sillitoe 1988: 6; Bouquet 2001).

The engaged agenda seeks to promote inclusion beyond development and museum contexts, to involve the wider discipline of socio-cultural anthropology. It is increasingly thought overdue and has the potential radically to alter research practice and approaches to teaching, and indeed what we think we know. It implies an overhaul of the entire research process so that it is more responsive to the concerns and requirements of communities, involving them as collaborators, from formulation of research project to the dissemination of findings (Co-operative Research Centre for Aboriginal and Tropical Health 2002; Denzin, Lincoln and Tuhiwai Smith 2008; Low and Merry 2010: S209). Many indigenous scholars argue that research should bring benefits to the community where carried out (Ka'ili 2012). In her chapter, Emma Cervone discusses the epistemological and methodological implications of such collaboration between engaged anthropologists and indigenous scholars, drawing on her Ecuador experiences. While indigenous academics tap into their communities' experience in formulating their research agendas, engaged anthropologists seek to further their aims by advancing participatory approaches. Building trust is a central issue regarding such engagement, by, for instance, demonstrably changing power relations within research projects, forging meaningful partnerships with indigenous representatives and scholars to gauge progress and access local concerns. In this volume we seek to further such journeying together through collaboration, forging necessary connections and promoting a positive exchange of ideas, expertise, and criticism.

While cooperation between engaged anthropology and indigenous research seems advantageous, indeed obligatory, particularly with the similarities between their philosophies, it is perplexing that they are working largely in isolation. There are considerable mutual misunderstandings given their markedly different views and approaches; most anthropologists ignoring indigenous scholars, who in their turn are largely antagonistic to outsider researchers. Let's break down these barriers. We may facilitate meaningful dialogue and exchange of ideas about future directions in research of, and by, local communities by bringing all parties together, seeking to draw on different cultural-intellectual traditions. What issues do indigenous studies see as central and what role is there for anthropological research? What is their perspective on the currently popular participatory/ engaged research paradigm? By having colleagues with different backgrounds – experience of indigenous universities, international agencies etc. – compare notes, experiences, ideas and so on, we seek parallels and learn lessons from one another. It is necessary frankly to compare and contrast approaches to documenting and furthering understanding of human ways of being in the world, exploring the advantages and disadvantages of different approaches to research, clearing up misconceptions and agreeing mutually beneficial future directions (in the spirit of Beck's 2004 'realistic cosmopolitanism').

Defining Indigenous Studies

What are indigenous studies? They feature research conducted by and for indigenous people rather than being about them (Co-operative Research Centre for Aboriginal and Tropical Health 2002; Center for World Indigenous Studies, n.d.). They have emerged from a deep-seated and understandable resentment with research that largely excludes communities, while being about them (Rigney 1999, 2006). They are similar to other subaltern demands through history for people to have more of a say in affairs that concern them, such as working class and suffragette movements. They feature not only research that aims to benefit the community, but also distinctive participatory methodologies advanced by indigenous academics, which build a transparent, genuinely two-way process, including democratic decision-making (Denzin, Lincoln, and Tuhiwai Smith 2008). Such as Kaupapa Māori research in Aotearoa (New Zealand), where indigenous people and scholars engage on equal terms (Tuhiwai Smith 1999: 183–95; Bishop 1996, 1999, 2005). In her chapter, Domenica Calabrò gives an insight into what collaborating as an anthropologist with Kaupapa researchers involves – 'where research meets Maori, or Maori meets research, on equalizing terms' (Tuhiwai Smith 1999: 185) – during her enquiries into the impact of the indigenization of New Zealand rugby on Māori identity.

What further distinguishes indigenous scholars from anthropologists? While defining the discipline of anthropology is a Sisyphean task – and hence not one I shall attempt here – it clearly differs from indigenous scholarship for those who engage in it. By definition, persons who identify as members of communities conduct indigenous studies there. This is arguably a significant distinction between indigenous scholars and anthropologists. Some think that to qualify as an anthropologist it is necessary to work in a different socio-cultural context – that is, traverse some socio-cultural boundary – because only in this way can you gain the necessary cross-cultural experience and awareness that is central to anthropology (Sillitoe 2007: 151). But most indigenous scholars have by default crossed such a socio-cultural Rubicon and so have an anthropological perspective, being exposed to Western education, language and philosophy – which in turn leads to further problems discussed below. Some even have an anthropological background, which they have presumably to keep in check, so as not overly to contaminate their indigenous credentials (for example Medicine 2001). Similarly, some Euro-Americans who only have experience of European society call themselves anthropologists and not sociologists, having formulated the idea of 'anthropology at home'. In the context of this volume, they presumably qualify as indigenous scholars too, which has a certain irony as many non-Western indigenous scholars would be offended to be labelled as anthropologists, whereas some of their Western counterparts fight to justify the use of the label for themselves, which suggests some further differences.

This contradiction relates to several long-standing anthropological conundrums, which result from trying to ride two horses at once – the culturally relative and comparative – that paradoxically seem to pull in opposite directions.[3] Indigenous studies focus by definition on a single region/community, and make no pretence at furthering understanding of all humankind. Those who engage in anthropological fieldwork likewise seek to further understanding of particular socio-cultural arrangements at certain historical periods, but whereas indigenous scholars do so with a view to furthering the interests of the local community, anthropologists do so to increase their knowledge of the range of such arrangements globally. They seek further to advance our (Western framed) understanding of the human condition from a comparative cross-cultural perspective, in which context it is arguably legitimate to employ the Western 'term of second intention' society. Such fieldwork, crossing into a different socio-cultural world, is arguably necessary not only to supply ethnographic data for cross-cultural intellectual hypothesizing but also to sharpen awareness of comparative issues, otherwise we remain rooted in our own ethnocentric world view, albeit we cannot entirely escape it, being subject to our own cultural conditioning as the postmodern critique affirms. While indigenous scholars, focussing on their communities, will further enrich our understanding of the range of human socio-cultural arrangements, and so supply grist to the comparative mill, they moreover promise to deepen the challenge to our understanding of what it is to be in the world, constructively confronting the postmodern view that has almost paralyzed social science (as science) for the past two decades or so.

Correctness of Term 'Indigenous'

The above definition of indigenous studies and its contrast with anthropology assumes that you buy into the idea of 'indigenous' as a category. Some social scientists criticise the idea of indigenous scholarship, finding the 'indigenous' adjective objectionable because it is difficult to define in these globalizing times with much migration, and potentially divisive and politically dangerous. There has been a furious debate recently over the correctness of using the word indigenous, detractors arguing that history shows that it is difficult to define, if not meaningless (Kuper 2003; Kendrick and Lewis 2004; Barnard 2006; Guenther et al. 2006). In his chapter, Dimitrios Theopopolus comments further on the debate. No one is truly indigenous to anywhere. There are 'no indigenous peoples. Every native began as an alien' (Beck 2004: 447). There is, for example, no such person as a thoroughly native Englishman because those who inhabit England are a mix of immigrants and invaders including Celts, Romans, Angles, Saxons, Vikings, Normans, with repeated influxes of Irish, Scottish, Jewish, Huguenot and more recently various Commonwealth and European citizens, resulting in a cosmopolitan if not mongrel culture – as captured in the Scottish bard's words 'England, dear England – the most colonized place on earth'.

---

3 I owe this vivid image to my friend Brian Morris.

There is anxiety from a liberal perspective that use of the term indigenous may encourage disagreeable xenophobic, even racial emotions. Also, unsurprisingly, some colonially founded states, such as the USA and Australia, do not wish to draw attention to indigenous issues and rights given their history of domination, ethnocide and even genocide (for example Wolfe 1994; Povinelli 1998; Battiste 2002; Four Arrows 2006). While some intellectuals warn against the use of the term indigenous, many people use it to describe themselves. Indeed attempts to discourage its use could arguably increase animosity, if persons perceive a threat to their identity or prerogative to consider themselves as native to somewhere – that is, belonging where they can relate to the values and way of life. Millions think of themselves, for instance, as English indigenes. This relates to that strand of social studies that concerns identity: the emotional need that humans evince to belong to a group, to a culture with its language and history. The chapter by Robyn Sandri, who identifies with the Aboriginal people of south Queensland, graphically illustrates the complexities of such concerns, particularly where indigenous communities have endured generations of colonial domination, including gross violation of human rights.

These identity concerns often involve an interest in tradition. We see this in the urge the Kwaio have to 'codify their customary law', mentioned in my opening comments. These interests in tradition are also subject to academic criticism. Archaeology and history show that all cultures change, albeit some apparently faster than others, and so an interest in traditional identity, thought of as unchanging custom passing much the same from generation to generation, is wrongheaded, for there is no such thing (Hobsbawm and Ranger 1983; Douglas 2004). While this may be correct from a Western intellectual perspective, it does not apparently stop some people – again often ethnic minorities of tribal origin – talking about tradition and their wish to keep it intact, the way they think it has always been for them. They express a need to protect it from extinction in the face of current global pressures and some even search for 'lost', often suppressed, traditions. They sometimes draw on the work of ethnographers to construct such a sense of identity and communicate it to others, as Dimitrios Theopopulos discusses in the context of the Embera speakers of Panama and their efforts to take advantage of the opportunities presented by cultural tourism. There is an intriguing contradiction here, in that regardless of the hostility some indigenous scholars may have for anthropologists, some draw heavily on earlier ethnographic research in asserting identities built on 'tradition' and 'custom'.

These efforts to reinstate tradition bring to mind the distinction between so-called hot and cold societies, or linear and circular ones. Criticism of the terms indigenous and traditional is of a piece with anthropologists' shift of interest away from their 'traditional' circular-tribal subject matter towards their own society's linear concepts and concerns with change. It is argued that while ideas of circular tradition may reflect the local view in some places, such a synchronic focus (with associated notions of structure and bogus ethnocentric present) is erroneous social science. It appears an overstatement, even misrepresentation, as occurs in many a radical change of perspective. Local ways, albeit subject to change, continue against the apparent odds. Hunter-gatherers continue to hunt and gather; shifting cultivators to shift and cultivate; communities to believe in local deities; clan obligations to inform social life etc. They are stubbornly resistant to the blandishments and threats of global capitalism and modernism, suggesting that there is something to peoples' insistence that tradition rules. It is possible that humankind will eventually develop a harmonious global culture where common approaches to issues will be the norm, but it is a long way off on current evidence, an aspect of which is the way people keenly defend their identities – as indigenous, traditional or whatever (see debate between Beck 2004 and Latour 2004).

Politics of Indigenity

The interest in local traditions suggests that not all are in a thrall at the prospect of change (often glossed optimistically as development) or globalization (for many, code for American world domination), which is ironically pushing some people in the opposite direction, seeking to reaffirm their indigenity and associated traditions, as they see what they value increasingly threatened by nominally democratic capitalism. Such views are expressed by George Dei, Priscilla Settee and Robyn Sandri in their chapters. They relate again to the political dimension of indigenous studies, as they attempt to overcome the legacy of domination and stereotyping by outsiders (Spivak 1988). The focus on custom supports efforts to establish legitimate entitlement to resources and gain some political leverage by substantiating traditional claims. We see this with some populations increasingly marking themselves off as native to certain regions as they fight for their rights and self-determination, of which national governments seek to dispossess them (see for instance the debate in *Anthropology Today* – Bowen 2000; Colchester 2002; McIntosh, Colchester and Bowen 2002; Rosengren 2002 – also Niezen 2003; Asch 2004). Many of those involved are tribal people, who have long been of interest to, even until recently largely defined anthropology, the only discipline to esteem them highly enough to wish properly and sympathetically to understand their ways of being. The entitlements of many tribal populations have been grossly violated from a liberal standpoint (as highlighted regularly by NGOs such as Survival International and Minority Rights Group, see for example the on-going struggle of the San people in Botswana (Minority Rights Group 2005; Survival International 2006; Oma and Thoma 2006)). They are trying to right historical wrongs and surely have a right to represent themselves as indigenous and have recourse to tradition, if they think it will help their interests?

But the political issues are complex, for instance who has the right to represent whom and what? The question of when, if ever, it is appropriate to involve oneself in the politics of communities to which one does not belong presents an on-going ethical dilemma for those who advocate engaged anthropology. The tussle is between 'becoming engaged in activism that seeks to reform features of social life to enhance social justice' and seeking to remain 'a disengaged outsider observing and recording social life' on the grounds that one has no 'ethical right to seek to change other ways of life' (Low and Merry 2010: S211–12). Many indigenous representatives argue that such politics is for local people. There are various indigenous movements, including locally rooted NGOs and educational bodies, currently seeking to voice their views internationally, sometimes holding alternative world summits (Sillitoe and Bicker 2004; Center for World Indigenous Studies n.d.). The chapter by Jayantha Perera discusses associated problems, such as peoples' claims to the right of self-determination when engulfed within nation states, which may pay lip service only to the ethic of free, prior and informed consent regarding activities that intrude into indigenous communities and trample on what these consider to be the good life. The politics of representation of knowledge is tricky for many indigenous activists: how representative are they of their communities? There is the thorny matter of who selects local representatives, and keeps an eye on their representation and so on. Often nobody, as they are self-appointed; few are democratically elected by their communities to represent their views. Whatever way you look at it, we remain one step removed from the 'real knowledge' that resides in local communities, and inevitably deal with representatives, albeit they may have a more direct, even authentic voice, as culture bearers. And they may more meaningfully foster participation, although on the other hand, they may not.

While the hoary question of the propriety of appearing to speak for others has been central to debates about anthropological involvement in development (Agrawal 1995, 1999; Hobart 1993; Shiva 1996; Escobar 1998), local representatives have to be cautious too of being accused of taking, altering and co-opting their community's knowledge to their own ends. There is a danger that communities may experience less outsider domination, only to find themselves subject to more insider domination. Local representatives are not neutral, after all, as there is no such thing as a value-free stance in social enquiries. They are as prone as any outsiders to impose their own moral views in research, and these may align only with those of a section of the local community. They may have vested interests that prompt them to represent views that are against the interests of some section of the community. The chances are that they will align with more powerful local factions and act in ways to maintain social hierarchies in their favour; which is an issue that Ray Nichol addresses in his chapter. The personal dilemmas of those who act simultaneously as 'insiders' and 'outsiders' can be considerable, finding themselves facing conflicting interests (Kaomea 2004; Jankie 2004).

Critiques of the use of the words 'indigenous' and 'traditional' appear to have more to do with Western intellectual preoccupations than those of people elsewhere who happily use these labels. Arguments against their use may be well intentioned, reducing the 'we' and 'them' distinctions that often characterize misuse of power, also possibly offensive racism, by encouraging us to consider ourselves all equally human with the same rights and so on, as Jayantha Perera argues. But they come perilously close to taking a colonial-like superior stance – 'we know better how dangerous such terms are at fomenting nasty sentiments, so stop using them', which is reminiscent of the 'we know better than you how you organize your social lives, having the idea of society that we deploy analytically'. It is unhelpful, even arguably ethnocentric, for privileged, globe-trotting, cosmopolitan academics to foist their concerns and worries on others, arguing that they should not talk in certain ways because it is divisive, especially when the concerns and worries of these populations are often pressing, involving basic human rights issues, and even on-going survival.

Criticism of the 'indigenous' adjective is even potentially more neo-colonial than these comments suggest (Low and Merry 2010: S212–13), in apparently denying that people should distinguish their socio-cultural traditions from others. It is understandable that some people suspect that it is cover to smother their way of being in the world as they choose. In effectively seeking to shut down debate, the anti-indigenous argument could be seen to justify the imposition of the views of the most powerful on the rest, on the grounds that there are no differences between us. And on one level this is so, as we are all human beings – the rub is: whose socio-culturally determined world view and understanding is going to dominate? The anti-view might be used, for example, to justify the modernization approach to development as it implies that this 'solution' should work everywhere (Sillitoe 2002: 133). And people will have society whether or not it jives with their view of social life because, denied any separate cultural identity, we can impose it on them. While few of those arguing against the 'indigenous' and 'traditional' adjectives would probably go along with this line of reasoning, the danger exists if we are unable to distinguish between communities in the manner these terms suggest, for otherwise the same views should logically apply everywhere.

If some people wish to speak of themselves as indigenous with respect to some cultural tradition, and can to their minds distinguish indigenous and traditional – categories that overlap considerably with such anthropological ones as society and culture – who has the authority to tell them otherwise? After all, even the UN, arbiter on global political issues, sanctions the use of the term indigenous with its Decade of the World's Indigenous People and concern for Indigenous Peoples' Rights (United Nations 1993; Minority Rights Group 2006). We need to get away from the ideology and back to the ethnography to further understanding of why and how people use such terms, and listen non-judgmentally to indigenous scholars.

Knowledge Differentials

Indigenous scholarship challenges not only the propriety of the categories we use but also, as mentioned above, Western ideas of what knowledge comprises, as Robyn Sandri and Emma Cervone point out in their chapters. Its perspectives may differ strikingly from those of social science, as illustrated by the absence, for instance, of the abstraction society as an organizing principle for explaining human behaviour. In contrast to such social philosophy, many indigenous scholars emphasize the moral, even religious dimensions of knowledge, which they believe involves learning what it is to be a socially responsible member of your community aware of your place in the wider environment, seeing the maturity of the self and achievement of wisdom as an on-going spiritual process (Manuelito 2005: 83; Ismail and Cazden 2005: 90; Barnhardt and Kawagley 2005; Universidad Intercultural Amawtay Wasi 2004). The chapter by George Dei illustrates the importance of the moral dimensions of education through a discussion of the use of traditional proverbs in some African communities. In this view, knowledge is not about having the wherewithal for explaining and doing things to the world but about understanding and being in it responsibly.

It is a worldview that is gradually gaining a hearing in the West, moving from the margins of wacky green spiritualism that claims supposedly Druidic roots, to the mainstream as we increasingly realize the need to respect nature and seek to reconnect with her. For instance, it is perhaps less likely today that we should witness a repeat of what occurred at the millennial conference of the Association of Social Anthropologists of the UK and Commonwealth devoted to discussing local knowledge in development (Sillitoe, Bicker and Pottier 2002). We invited to the meeting some advocates of the endogenous development approach (COMPAS), which seeks to overcome the materialistic bias of Western attitudes by taking people's spiritual and social well-being as well as their material well-being into account, who to our embarrassment encountered academic arrogance, indeed rudeness, of the worst kind, with many delegates dismissing or talking down to them.

The role of knowledge in contemporary society may be different in the eyes of indigenous scholars, as is evident in attitudes to rights over it. The Western approach stresses the ownership of knowledge with patents and so on to protect it and ensure its privileged exploitation, often in some productive process for material gain. Other approaches are more open and treat knowledge as common property, something to which the community has collective rights, to be shared to the benefit of all (Posey, Dutfield and Plenderleith 1995; Brush and Stabinsky 1996; Dutfield 1999; Clift 2007). These differences between knowledge traditions prompt some indigenous scholars to speak of the need for a clean break, seeing no hope of reconciliation, only continuing domination, as exemplified in the Western approach to controlling knowledge and manifest starkly in bio-prospecting and piracy experienced by some communities. It is all of a piece with indigenous scholars seeking to break away from outside power structures that have undermined their epistemologies, an aspect of the aforementioned struggles for self-determination, securing human and land rights, and even cultural survival (McCarty et al. 2005; Tuhiwai Smith 2005: 95). In this context some speak of decolonizing their knowledge (Tuhiwai Smith 1999; Kagendo and Swadener 2004), which is a theme that Emma Cervone addresses in her chapter. How feasible is this aim?

The Literacy Conundrum and Other Representations

Again, playing the devil's advocate, it is arguable that written Western documentation and discussion inevitably dominated where Europeans came across oral traditions in the Americas, Africa and Australasia. After all, a literate tradition leaves an enduring record whereas spoken words do not. Indeed I think that a good case can be made in support of many ethnographers' efforts to record peoples' ways of life in these regions before they were overwhelmed and changed by colonial and subsequently capitalist globalizing forces. They esteemed these cultures highly enough to spend their careers documenting and furthering sympathetic understanding of them. Their writings comprise a significant record of cultural heritage and history that would otherwise be lost to humanity, which as pointed out above many indigenous people today draw upon in reconstructing identities more authentic to their way of being in the world and defending their right to do so. The astonishing Smithsonian records come to mind here, a resource mined by First Nations people in search of evidence about their ancestors' ways, albeit largely ignored currently by anthropologists in their cross-cultural attempts to further our understanding of humanity (see, for example, Dorsey 1884; Cushing 1896; Mooney 1896; Fletcher and La Flesche 1911; Boas 1921; Radin 1923; Swanton 1928; Bunzel 1932).

But, of course, Western intellectual traditions even came to dominate where people had their own literary heritage, such as in South Asia, China and the Arab world – under the banner of Oriental studies besides anthropology. Although some anthropologists may refer to this literature, particularly those working in the 'Orient', the ideas it contains somehow fail much to inform their 'theories'. For instance, it is only recently that I have learned about the work of Ibn Khaldun, who is probably one of the world's first sociologists, writing on social issues some hundreds of years before European social philosophy hit upon them (Gellner 1975; Alatas 2006). His concept of *asabiyyah* foreshadows ideas of social solidarity that did not appear in Europe until the nineteenth century (Durkheim 1933), and Ibn Khaldun also advanced a theory of a rural–urban cycle of social change that foreshadows twentieth-century ideas of rural–urban continua (Redfield 1947). Moreover, through the assiduous efforts of colonial, missionary and development agencies, there are now literate people everywhere capable of documenting and discussing for themselves their particular culturally informed views and understanding – many increasingly styling themselves indigenous scholars – and yet Western academic traditions continue to dominate around the globe (Ribeiro and Escobar 2006).

Literacy brings other problems for indigenous scholars (Tuhiwai Smith 1999: 35–7), namely inevitable pollution by the Euro-American intellectual tradition, when they are caught up in its formal education system, as Rachel Shah points out in discussing the problems that people in Papua encounter with formal schooling. The challenge indigenous academics face in this event, is distancing themselves from the indoctrination they have received from a Western education to find the authentic voice of their people. This relates to above comments about indigenous scholars being anthropologists by default – whatever they may think of anthropologists – in the sense that their exposure to Western education and thought means that they have an awareness of two entirely different socio-cultural traditions and wrestle with the implications. It suggests that we need a reverse anthropology; that is, one that looks at the West through an Aboriginal or Amazonian philosophic-cultural lens. To what extent is this possible or do exchanges such as in this volume simply perpetuate structures of intellectual domination, taking place in a European educational context and following its rules of debate?

Looked at this way, both insider and outsider scholars face the same fundamental problem, but seen from opposite sides: this suggests that collaboration or 'interweaving' (Ka'ili 2012) should be fruitful, the two perspectives complementing one another. Portraying non-Western knowledge in ways understandable to those socialized into the Euro-American tradition is one of anthropology's defining objectives. It has long wrestled with the problem of faithfully representing other lifeways and worldviews, especially using Western categories and methods of documentation – such as the idea of society, imposed as if a term of first intention. It is not the only understanding possible or legitimate, and certainly not the one by which we should necessarily frame our understanding of particular socio-cultural traditions. The extent to which it is possible to escape our intellectual frame of reference has become a central topic of debate, particularly with the advent of postmodernism, which argues that anthropology's ideological (or theoretical) concerns inevitably distort its subject matter. It is likely that jointly exploring and reviewing indigenous critiques will help address this problem; the indigenous scholar an integral party, making us aware of issues and helping us to understand other views (Center for World Indigenous Studies n.d.). Arguably this is truer to anthropology, allowing culture-bearers to drive research enquiries, and by deferring to indigenous scholars we may potentially have a way (at least partially) to get beyond the postmodern hermeneutic impasse.

In what ways do such scholars' presentations of other 'ways of being in the world' differ (Ribeiro and Escobar 2006)? They are often fiercely critical of outsider anthropologists' representations – such as my frustrated University of Papua New Guinea colleague – who perceive the discipline to be wandering lost in a jargon jungle impenetrable to them, with preoccupations that seem unrelated to local views and concerns (Owusu 1978; Nakata 1998). Again, it is an aspect of intellectual control and power, for those unable to comprehend what is going on are inevitably shut out. It is necessary for anthropology to take their criticisms and views seriously. How credible would foreigner written travel guidebooks be, if what they contained was unrecognizable to locals who read them? But do indigenous scholars represent their place and community more faithfully? They will surely distort the local perspective to some extent, particularly if they seek to meet Western academic expectations. These challenges relate again to the political issues of representation discussed above. The question of how the views of indigenous scholars differ from those of the people they seek to represent is as crucial as differences between them and non-indigenous researchers.

Challenges of Indigenous Education

This brings us to education, which is emerging as central to indigenous studies (Villegas, Neugebauer and Venegas 2008), as demonstrated by the contributions of Priscilla Settee, Ray Nichol and Rachel Shah to this book. Formal Western style education may paradoxically both erode indigenous knowledge and possibly uphold it, if appropriately organized – and much debate revolves around appropriateness (Bates et al. 2009; Wangoola 2012). The establishment of indigenously framed education programmes epitomizes attempts to break free of distorting Western intellectual views and dominance and to wrestle back control of the representation and reproduction of local ways; for instance, the Mpambo Afrikan Multiversity that promotes African scholarship and knowledge using local languages, challenging mainstream academic thinking (Wangoola 2012). Regarding universities, for instance, there are calls for those that seek to include indigenous students, to embrace their worldviews, not expect them somehow to put aside their cultural heritage (Mihesuah and Wilson 2004; Kuokkanen 2007). In her chapter, Priscilla Settee, who intriguingly is a member of the North American Cree Nation working with a South American indigenous university in Peru, gives a first-hand account of the problems that such students face and the implications, for instance with respect to motivation and retention issues.

The emergence of indigenous educational institutions to transmit and represent local tradition in culturally appropriate ways clearly relates to concerns mentioned earlier over identity and self-representation. An array of institutions has appeared, from tradition schools to indigenous universities, variously supported by local NGOs and governments, which seek to work in culturally appropriate ways, offering a range of courses and workshops tailored to local custom. Teachers with native credentials seek to instruct students from their communities in culturally apt ways, drawing on indigenous epistemologies to organize syllabuses, courses, and research. Several different parallel approaches to indigenous teaching and research have emerged, struggling with similar questions but formulating diverse locally appropriate answers. Some of these occur in the formal education sector, where enlightened educationalists strive for indigenously appropriate approaches, such as Ray Nichol proposes in his chapter, drawing on a long career as an educationalist seeking ways forwards in Australasian contexts. But elsewhere, authorities seek to impose foreign educational structures, often deeply flawed and poorly staffed, as Rachel Shah describes for Papua, where schooling is part of an assimilation policy that aims to turn Papuans into good Indonesian citizens, and threatens to repeat mistakes made elsewhere.

These fledgling indigenous initiatives and institutions face considerable problems, not least of recognition. To what extent is it possible, or indeed necessary, to reconcile indigenous teaching and enquiry with mainstream education? How do their respective pedagogic philosophies converge and diverge, the indigenous side featuring a range of worldviews that contrast and sometimes conflict with Enlightenment principles of rational enquiry, and what is the significance of these differences? This returns us to the issue of domination again, as Ray Nichol reminds us in his chapter. There is a tendency for Western-style education to assume that it is superior, largely because of its technological associations. Its hegemony derives from the application of certain knowledge, which has resulted in awesome industrial capacity that allows humans unprecedented ability to intervene in the world. This apprises efforts to interface indigenous with scientific knowledge in development contexts, namely to facilitate the up-take of assumed technological benefits, by representing local knowledge and practices more effectively to development agencies, winning respectability so that they can play their part in this process (Sillitoe, Dixon and Barr 2005).

These efforts assume that local knowledge and practices have something to contribute more generally too. Critics of such techno-capitalist framed development are increasingly asking if it is necessarily a good thing, pointing out that we do not seem to have the wisdom (and certainly not the political structures) to handle contemporary technological knowledge, as ever more evident with growing concerns over damage to the natural environment, notably climate change, and horrendously destructive wars. And there is certainly, of course, more to life than material standard of living, as we are increasingly realizing with a growing focus on all aspects of well-being, including the social and emotional. These are certainly themes that resonate with indigenous education initiatives that seek to find space for alternative views of how to be in the world, as George Dei's chapter stresses.

The challenges facing indigenous education look different when viewed from the mainstream and indigenous sides. Indigenous scholarship can prompt strong mainstream criticism. Some think that it threatens to compromise accepted intellectual standards and academic rigour. This view is evident in the comment quoted earlier, asserting social scientists' authoritative use of the society concept. These worries relate to loss of disciplinary authority and control over the quality of teaching and research. While indigenous scholarship may question Western intellectual pre-occupations, we should surely welcome this as good anthropology, with culture-bearers – that is, the ethnography – leading enquiries, which is in line with my tongue-in-cheek call for ethnographic-determinism (Sillitoe 2003: 336; 2010). Furthermore, is it not possible that we might agree some universal standards for indigenous scholarship and anthropological research?

But we should not underestimate the problems that we face if these standards conflict with the powerful metropolitan educational establishment – for anthropology is only a tiny and politically weak discipline. The career implications for academics seeking to synchronize their work sympathetically with indigenous scholarship are considerable (especially early in their careers) with pressures to conform to disciplinary standards of 'true' scholarship (Mahmood 2012), exacerbated currently with RAE/REF[4] demands imposed by accountancy-driven

---

4   The REF is the 'Research Excellence Framework', which is the successor of the RAE 'Research Assessment Exercise', which are the UK government's mechanisms for assessing the worth of university research and funding.

Western culture (Shore and Wright 1999). These inhibit intellectual freedom and true experimentation. It is arguably foolhardy to allow participation to drive the research agenda with these constraints (Low and Merry 2010: S213). Allowing indigenous scholarship a lead role may prevent researchers focusing on topics they are knowledgeable of, which, with the trend for ever narrower specialization within academia, is another problem. In effect it may seem to deny anthropologists any expertise (Sillitoe 2007: 154–5). Certainly they may no longer be free to contribute to the ideological (theoretical) debate of the moment as they choose, to advance their careers, an intriguing prospect as mentioned above.

The challenges viewed from the indigenous side concern the struggle to get Western education to accept it as reputable. This is evident for instance in the difficulties that indigenous scholars face in publishing their work in pukka Euro-American outlets (such as academic journals, university presses etc.) because not deemed to be of sufficient intellectual merit. It illustrates again the puzzling aforementioned intellectual arrogance, for a discipline that purports to further understanding of other socio-cultural ways to deny the voice of scholar-culture-bearers. One indigenous response is: why bother anyway? The costs for indigenous educational institutions of winning a certain level of accreditation by mainstream educational authorities may be too high, recognition demanding too many compromises in their representation and transmission of indigenous philosophies, often tacit and experiential, inimical to structuring along the formal lines of Western institutions, even radically styled ones. Indeed this is arguably an issue for anthropology generally too – see below. Instead, indigenous scholars should seek to have their knowledge accepted on its terms, just as the West accepts Asian transcendental meditation as an alternative knowledge tradition, acknowledging that if you meditate you achieve a different state of consciousness and understanding of being in the world.

Independence may have resource implications – although some may argue that if you wish to do things locally and maintain local control, you should resource activities locally. For instance, one response to the publication problem is for indigenous scholars to create their own outlets and perhaps show up Western academics with their inclusive ethos – so often in my experience an indigenous value (that is, not wishing to dominate or be subordinate). But they may have considerable difficulty funding themselves. This has certainly been our experience, establishing two journals in Bangladesh – *Grassroots Voice* and *Local Voice* – devoted to local knowledge in development issues (Sen, Angell and Miles 2000: 214). More positively, the emergence of electronic publishing may afford more affordable opportunities. But there is a danger that any such indigenous alternatives will be thought inferior; certainly my colleagues have judged papers that I have published in these two Bangla journals as not REFerable. Also, of course, this go-it-alone response defeats the objective of this volume, to bring all sides together on a journey to explore collaborative possibilities. In these respects, the interface between indigenous and Western educational institutions merits further discussion.

Indigenous Education: Implications for Anthropology

The establishment of indigenous educational and research institutions is likely to have considerable repercussions for anthropology. The educational implications of harmonizing informal indigenous pedagogy with formal Western education are potentially profound for the discipline, if not other subjects too. In future, anthropological education should arguably expect to follow the indigenous way by and large; the views and experiences of native scholars coming to inform teaching about their cultures in Western institutions. They have the potential to revolutionize not only the way we teach but also what we teach our students; in short the entire anthropological approach to education.

In what ways do the approaches and methods of indigenous educational institutions differ in representing local knowledge and practice: are they more authentic and valid (for example Swisher 1998; Tuhiwai Smith 1999) and what does anthropology have to learn from them? If we decide they are more authentic and valid, then we have logically to follow them. It seems only right that indigenous institutions and their staff should inform teaching and research on their cultures, entering into a constructive dialogue with anthropologists to forge mutually helpful scholarly links. What are the implications for Western education, of learning the indigenous way – which is often tacit and experiential – frequently eliciting and documenting knowledge differently, even differing over what constitutes knowledge? It implies new approaches to education and the need to consider the prospect of having to think about learning anthropology in novel ways.

It is intriguing to contemplate the implications of teaching students about Maori or Aboriginal or any other cultural reality in terms determined by culture-bearer-scholars and not Western-social-scientists. For instance, imagine seminars where it is necessary to undergo aspects of Aranda initiation to understand human–environment relations the Desert Aborigine way, or to imbibe the *ebene* hallucinogen of the Yanomamo to appreciate shamanism after an Amazonian Indian, or learn to trance dance as a San !Kung to connect with aspects of life in the Kalahari desert. They imply different educational standards. They may, for instance, give a novel spin to knowledge ownership, mentioned earlier; if we learn about the Dreamtime like an Aranda, for example, this may impose restrictions on sharing the knowledge that is kept secret between initiates (see Allen's 1998 interesting discussion of the pedagogic implications of Laguna Pueblo secrecy). The consequences for examinations are fascinating to contemplate and intimate the sort of revolution implied in education: away from tick-box pedagogic-factories turning out transferable-skilled graduates for the job market; a revolution that increasing numbers of critics think necessary (Mahmood 2012), albeit perhaps not in this way.

The variety of approaches to education presents further conundrums. We should not expect, of course, as anthropologists, to come up with any unified global way of representing, teaching and 'preserving' socio-cultural knowledge. And we need to break with the idea of Western knowledge versus all others. There is a political aspect here again relating to points made above, with approaches varying from region to region depending not only on culture but also history (of colonization etc.). This is one of the key messages of indigenous knowledge research in development, namely that the search by agencies for one development model or solution is inappropriate, as the same approach will not apply everywhere (Sillitoe, Dixon and Barr 2005: 22). It is necessary to acknowledge the plurality of approaches, comparing and contrasting the experiences and insights of indigenous educationalists and researchers from various parts of the world.

But what are the implications for ensuring the quality and rigour of teaching and research, points made above? Again we find that the indigenous scholar approach, allowing for a multitude of views, flies in the face of current policy in Western higher education that seeks to standardize everything according to checklists of 'learning outcomes', 'key skills', 'module proformas' 'methods outlines' etc. Indigenous scholarship may help combat the consequent danger of intellectual ossification. While from a Western intellectual viewpoint, a good knowledge of various standard approaches to a subject is necessary to undertaking sound enquiries, we risk placing too much emphasis on these and producing research robots who will reproduce the methodological recipe everywhere, when we need to imbue a healthy element of flexibility and innovativeness in our approach to teaching and research, in order to allow other views of what knowledge is and being in the world to emerge and inform work in the best anthropological tradition. It is necessary to further understanding of how educational issues influence indigenous institutions, such as the impact of national policies and programmes on local communities, as Rachel Shah describes for Papua, and the need for these to vary as necessary from place to place. Otherwise we are going to go on imposing our view of society on everyone.

Threats to Anthropology?

In seeking to grant analytical space to alternative 'terms of second intention', such as ideas of 'society' or otherwise, anthropology may run the risk of falling further apart, when it is already, as noted, a difficult enough subject to encompass and define (Sillitoe 2007: 149–51). It is arguable that supporting indigenous efforts to develop enquiries and teaching appropriate and beneficial to local communities, which strikes those of us advocating a more engaged anthropology as ethically the right way to go, may potentially contribute to the discipline's further atomization. This is possibly another reason, largely unspoken and perhaps even unconscious, for resistance to allowing a larger voice to other cultural views, such as via indigenous scholarship.

The foregoing discussion of the definition of indigenous studies touched on the problems met in demarcating the diffuse discipline of anthropology, in seeking to distinguish anthropologists from sociologists and asking where indigenous scholars stand in relation to the two, noting the challenges of negotiating deep cultural-linguistic differences. These problems relate in turn to the tendency of anthropologists to subscribe to the theory-cum-ideology of the moment, which keeps their wayward disciplinary flock together in some measure, and we can see the danger they may sense in allowing the multifarious voices of indigenous scholarship more prominence, in threatening to undo this connection – in so far as current theory-cum-ideology does not call directly to issues of concern to local communities. The argument is not that social scientists should cease to use their disciplines' concepts to further their understanding of human behaviour, but rather that they need to act on the postmodern acknowledgement that they are peculiar to a certain cultural and historical moment, and allow more scope for those who supply cross-cultural behavioural grist to their mill, directly to express their understandings.

The dangers are surely worth courting for anthropology, the central tenet of which is that we have much to learn from other cultural ways of understanding how we are in the world. These are intellectually exciting and challenging times. And they may not be that threatening anyway. Firstly, the ideology of much indigenous scholarship is reconciliatory and inclusive, not dominating and exclusive. Let's walk the path together. Secondly, the cross-cultural comparative tradition of anthropology continues to give it identity, Western intellectuals, as noted earlier, using their knowledge – albeit maybe distorted – of others' ways, to further their understanding of the human socio-cultural condition globally. We think comparison useful and that we may learn from each other's experiences; albeit those working comparatively will have to work harder to define the terms they use to label any range of institutions or behaviours from different cultures, the indigenous tradition resulting in each being represented according to its own 'terms of second intention' and not pre-packaged into ones – such as society or clan, lineage, exchange, chief or whatever – ready for comparative debate. The positive outcome will be to keep the primary ethnography more clearly distinguished from secondary intellectual activities: a goal worth striving for in itself.

The largest threat to anthropology is not to engage with these challenges, and for indigenous scholars to march on alone.

References

Agrawal, A. 1995. Dismantling the divide between indigenous and scientific knowledge. *Development and Change*, 26, 413–39.

—— 1999. On power and indigenous knowledge, in *Cultural and Spiritual Values of Biodiversity*, edited by D.A. Posey. Nairobi: United Nations Environment Programme; and London: Intermediate Technology Publications, 177–80.

Alatas, S.F. 2006. A Khaldunian exemplar for a historical sociology for the South. *Current Sociology*, 54(3), 397–411.

Allen, P.G. 1998. Special problems in teaching Leslie Marmon Silko's ceremony, in *Natives and Academics: Researching and Writing about American Indians*, edited by D. Mihesuah. Lincoln: University of Nebraska Press, 55–64.

Antweiler, C. 1998. Local knowledge and local knowing: An anthropological analysis of contested 'cultural products' in the context of development. *Anthropos*, 93, 469–94.

Asch, M. 2004. Political theory and the rights of indigenous peoples. *Canadian Journal of Sociology*, 29, 150–52.

Banerjee, S.B. and Linstead, S. 2004. Masking subversion: Neocolonial embeddedness in anthropological accounts of indigenous management. *Human Relations*, 57(2), 221–47.

Barnard, A. 2006. Kalahari revisionism, Vienna, and the 'indigenous peoples' debate. *Social Anthropology*, 14(1), 1–16.

Barnhardt, R. and Kawagley, A. 2005. Indigenous knowledge systems and Alaska native ways of knowing. *Anthropology and Education Quarterly*, 36(1), 8–23.

Bates, P., Chiba, M., Kube, S. and Nakashima, D. (eds) 2009. *Learning and Knowing in Indigenous Societies Today*. Paris: UNESCO.

Battiste, M. (ed.) 2002. *Reclaiming Indigenous Voice and Vision*. Vancouver: UBC Press.

Beck, U. 2004. The truth of others: A cosmopolitan approach. [Translated by Patrick Camiller.] *Common Knowledge*, 10(3), 430–49.

Bishop, R. 1996. *Collaborative Research Stories: Whakawhanaungatanga*. Palmerston North: Dunmore Press.

—— 1999. Kaupapa Maori research: An indigenous approach to creating knowledge, in *Maori and Psychology: Research and Practice – The Proceedings of a Symposium Sponsored by the Maori and Psychology Research Unit*, edited by N. Robertson. Hamilton: Maori and Psychology Research Unit, 1–6.

—— 2005. Freeing ourselves from colonial domination in research: A Kaupapa Maori approach to creating knowledge, in *The Sage Handbook of Qualitative Research*, edited by N. Denzin and Y. Lincoln. Thousand Oaks, CA: Sage, 109–38.

Boas, F. 1921. Ethnology of the Kwakiutl (based on data collected by George Hunt). *Thirty-fifth Annual Report of the Bureau of American Ethnology to the Secretary of the Smithsonian Institution, 1913–14* (Parts 1 and 2, 43–1473). Washington: Government Printing Office.

Bouquet, M. (ed.) 2001. *Academic Anthropology and the Museum: Back to the Future*. New York: Berghahn Books.

Bowen, J.R. 2000. Should we have a universal concept of 'indigenous peoples' rights? Ethnicity and essentialism in the twenty-first century. *Anthropology Today*, 16(4), 12–16.

Brush, S. and Stabinsky, D. 1996. *Valuing Local Knowledge: Indigenous People and Intellectual Property Rights*. Washington, DC: Island Press.

Bunzel, R.L. 1932. Zuñi ritual poetry. *Forty-seventh Annual Report of the Bureau of American Ethnology to the Secretary of the Smithsonian Institution, 1929– 30.* Washington: Government Printing Office, 611–835.

Center for World Indigenous Studies n.d. Source: http://asp.cwis.org/about.htm, accessed: 9 October 2011.

Cervone, E. 2007. Building engagment: Ethnography and indigenous communities today. *Transforming Anthropology*, 15(2), 97–110.

Chambers, R. 1997. *Whose Reality Counts?: Putting the First Last.* London: Intermediate Technology.

Clift, C. 2007. Is intellectual property protection a good idea? in *Local Science vs. Global Science: Approaches to Indigenous Knowledge in International Development*, edited by P. Sillitoe. Oxford: Berghahn, 191–207.

Colchester, M. 2002. Indigenous rights and the collective conscious. *Anthropology Today*, 18(1), 1–3.

Cooke, B. and Kothari, U. (eds) 2001. *Participation: The New Tyranny?* London: Zed Books.

Co-operative Research Centre for Aboriginal and Tropical Health 2002. *Indigenous Research Reform Agenda: Rethinking Research Methodologies.* Casuarina (Northern Territory): Co-operative Research Centre for Aboriginal and Tropical Health.

Cushing, F.H. 1896. Outlines of Zuñi creation myths. *Thirteenth Annual Report of the Bureau of Ethnology to the Secretary of the Smithsonian Institution, 1891–92.* Washington: Government Printing Office, 321–447.

Denzin, N.K., Lincoln, Y. and Tuhiwai Smith, L. (eds) 2008. *Handbook of Critical and Indigenous Methodologies.* London: Sage.

Dorsey, Rev. J.O. 1884. Omaha sociology. *Third Annual Report of the Bureau of Ethnology to the Secretary of the Smithsonian Institution, 1881–82.* Washington: Government Printing Office, 205–370.

Douglas, M. 2004. Traditional culture – let's hear no more about it, in *Culture and Public Action*, edited by V. Rao and M. Walton. Stanford: Stanford University Press, 85–109.

Durkheim, E. 1933 [1893]. Division of labour in society. [Translated by G. Simpson.] London: Macmillan.

Dutfield, G. 1999. The public and private domains: intellectual property rights in traditional ecological knowledge. *Electronic Journal of Intellectual Property Rights*, Oxford IP Research Centre Working Paper 03/99. Source: http://www.oiprc.ox.ac.uk/EJWP0399.html, accessed: 3 February 2012.

Ellen, R. and Harris, H. 2000. Introduction, in *Indigenous Environmental Knowledge and its Transformations*, edited by R. Ellen, P. Parkes and A. Bicker. Amsterdam: Harwood Academic, 1–33.

Emery, A.R. 2000. *Integrating Indigenous Knowledge in Project Planning and Implementation.* Nepean (Ontario): Partnership Publication with Kivu Nature Inc., The International Labour Organization, The World Bank and Canadian International Development Agency.

Encyclopædia Britannica 2009. Intention. Source: http://www.britannica.com/EBchecked/topic/289881/intention, accessed: 12 November 2011.

Escobar, A. 1998. Whose knowledge, whose nature? Biodiversity, conservation and the political ecology of social movements. *Journal of Political Ecology*, 5, 53–82.

Flaherty, M.G., Denzin, N.K., Manning, P.K. and Snow, D.A. 2002. Review symposium: Crisis of representation. *Journal of Contemporary Ethnography*, 31(4), 478–516.

Fletcher, A.C. and La Flesche, F. 1911. Te Omaha tribe. *Twenty-seventh Annual Report of the Bureau of American Ethnology to the Secretary of the Smithsonian Institution, 1905–6*. Washington: Government Printing Office, 17–654.

Four Arrows (Don Trent Jacobs) (ed.) 2006. *Unlearning the Language of Conquest Scholars Expose Anti-Indianism in America*. University of Texas Press.

Gellner, E. 1975. Cohesion and identity: The Maghreb from Ibn Khaldun to Emile Durkheim. *Government and Opposition*, 10(2), 203–18.

Greene, S. 2004. Indigenous people incorporated? Culture as politics, culture as property in pharmaceutical bioprospecting. *Current Anthropology*, 45, 211–37.

Grenier, L. 1998. *Working with Indigenous Knowledge. A Guide for Researchers*. Ottawa: IDRC.

Guenther, M., Kendrick, J., Kuper, A., Plaice, E., Thuen, T., Wolfe, P., Zips, W. and Barnard, A. 2006. The concept of indigeneity. *Social Anthropology*, 14(1), 17–32.

Hendry, J. 2002. Being ourselves for us: Some transformative ideas of ethnographic display. *Journal of Museum Ethnography*, 14, 24–37.

Herle, A. with Bani, M. 1998. Research notes: Collaborative projects on Torres Strait collections. *Journal of Museum Ethnography*, 10, 115–29.

Hobart, M. 1993. Introduction: The growth of ignorance? in *An Anthropological Critique of Development: the Growth of Ignorance*, edited by M. Hobart. London: Routledge, 1–30.

Hobsbawm, E. and Ranger, T.O. (eds) 1983. *The Invention of Tradition*. Cambridge: Cambridge University Press.

Ismail, S. and Cazden, C. 2005. Struggles for indigenous education and self-determination: Culture, context, and collaboration. *Anthropology and Education Quarterly*, 36(1), 88–92.

Jacob, P. 2003. Intentionality. *Stanford Encyclopedia of Philosophy*. Source: http://plato.stanford.edu/entries/intentionality/, accessed: 13 May 2012

Jankie, D. 2004. 'Tell me who you are': Problematizing the construction and positionalities of 'insider'/'outsider' of a native ethnographer in postcolonial context, in *Decolonizing Research in Cross-Cultural Contexts: Critical Personal Narratives*, edited by M. Kagendo and B.B. Swadener. New York: State University of New York Press, 87–106.

Kagendo, M. and Swadener, B.B. (eds) 2004. *Decolonizing Research in Cross-Cultural Contexts: Critical Personal Narratives*. New York: State University of New York Press.

Ka'ili, T.O. 2012. Felavai, interweaving indigeneity and anthropology: The era of indigenising anthropology, in *Anthropologists, Indigenous Scholars and the Research Endeavour: Seeking Bridges Towards Mutual Respect*, edited by J. Hendry and L. Fitznor. Abingdon: Routledge, 21–7.

Kaomea, J. 2004. Dilemmas of an indigenous academic: A native Hawaiian story, in *Decolonizing Research in Cross-Cultural Contexts: Critical Personal Narratives*, edited by M. Kagendo and B.B. Swadener. New York: State University of New York Press, 27–44.

Keesing, R.M. 1992. *Custom and Confrontation: The Kwaio Struggle for Cultural Autonomy*. Chicago: Chicago University Press.

Kendrick, J. and Lewis, J. 2004. Indigenous peoples' rights and the politics of the term 'indigenous'. *Anthropology Today*, 20(2), 4–9.

Kuokkanen, R. 2007. *Reshaping the University: Responsibility, Indigenous Epistemes and the Logic of the Gift*. Vancouver: University of British Columbia Press.

Kuper, A. 2003. The return of the native. *Current Anthropology*, 44, 389–402.

Lamphere, L. 2003. The perils and prospects for an engaged anthropology: A view from the United States. *Social Anthropology*, 11(2), 153–68.

Lassiter, L.E. 2005. *The Chicago Guide to Collaborative Ethnography*. Chicago: University of Chicago Press.

Latour, B. 2004. Whose cosmos, which cosmopolitics? Comments on the peace terms of Ulrich Beck. *Common Knowledge*, 10(3), 450–62.

Low, S.M. and Merry, S.E. 2010. Engaged anthropology: Diversity and dilemmas: An introduction to Supplement 2. *Current Anthropology*, 51, No. S2: S203–S226.

McCarty, T., Borgoiakova, T., Gilmore, P., Lomawaima, K. and Romero, M. 2005. Indigenous epistemologies and education – Self-determination, anthropology, and human rights. *Anthropology and Education Quarterly*, 36(1), 1–7.

McIntosh, I., Colchester, M. and Bowen, J.R. 2002. Defining oneself, and being defined as, Indigenous. *Anthropology Today*, 18(3), 23–5.

Mahmood, C.K. 2012. A hobby no more: Anxieties of engaged anthropology at the heart of empire. *Anthropology Today*, 28(4), 22–5.

Malinowski, B. 1922. *Argonauts of the Western Pacific*. London: Routledge and Kegan Paul.

Manuelito, K. 2005. The role of education in American Indian self-determination: Lessons from the Ramah Navajo Community School. *Anthropology and Education Quarterly*, 36(1), 73–87.

Medicine, B. (with S-E. Jacobs, ed.) 2001. *Learning to be an Anthropologist and Remaining 'Native': Selected Writings*. Champaign: University of Illinois Press.

Mihesuah, D. and Cavender Wilson, A. (eds) 2004. *Indigenizing the Academy: Transforming Scholarship and Empowering Communities*. Lincoln: University of Nebraska Press.

Minority Rights Group 2005. Botswana must amend discriminatory tribal policies, states UN Committee. Source: http://www.minorityrights.org/news_detail. asp?ID=351, accessed: 17 April 2012.

Minority Rights Group 2006. UN makes move for indigenous peoples' rights: MRG urges support of General Assembly. Source: http://www.minorityrights. org/TempNewsArticles/UN_HRC_Statement.html, accessed: 17 April 2012.

Mooney, J. 1896. The ghost-dance religion and the Sioux outbreak of 1890. *Fourteenth Annual Report of the Bureau of Ethnology to the Secretary of the Smithsonian Institution, 1892–93*. Washington: Government Printing Office, 641–1110.

Morauta, L. 1979. Indigenous anthropology in Papua New Guinea [and Comments and Reply]. *Current Anthropology* 20(3), 561–76.

Mosse, D. 2005. *Cultivating Development: An Ethnography of Aid Policy and Practice*. London: Pluto.

Nakata, M. 1998. Anthropological texts and indigenous standpoints. *Journal of Aboriginal Studies*, 2, 3–12.

Niezen, R. 2003. *The Origins of Indigenism. Human Rights and the Politics of Identity.* Berkeley: University of California Press.

Oma, K.M. and Thoma, A. 2006. Indigenous San knowledge and survival strategies, in *Indigenous Peoples' Wisdom and Power: Affirming our Knowledge Through Narratives*, edited by J.E. Kunnie and N.I. Goduka. Aldershot: Ashgate, 3–18.

Owusu, M. 1978. Ethnography of Africa: The usefulness of the useless. *American Anthropologist*, 80, 310–34.

Peers, L. and Brown, A.K. (eds) 2003. *Museums and Source Communities: A Routledge Reader*. New York: Routledge.

Posey, D.A., Dutfield, G. and Plenderleith, K. 1995. Collaborative research and intellectual property rights. *Biodiversity and Conservation*, 4, 892–902.

Povinelli, E.A. 1998. The state of shame: Australian multiculturalism and the crisis of indigenous citizenship. *Critical Inquiry*, 24 (winter), 575–610.

Radin, P. 1923. The Winnebago tribe. *Thirty-seventh Annual Report of the Bureau of American Ethnology to the Secretary of the Smithsonian Institution, 1915–16*. Washington: Government Printing Office, 35–550.

Redfield, R. 1947. The folk society. *American Journal of Sociology*, 52, 293–308.

Ribeiro, G.L. and Escobar, A. (eds) 2006. *World Anthropologies: Disciplinary Transformations in Systems of Power.* Oxford: Berg.

Rigney, L.I. 1999. Internationalisation of an indigenous anti-colonial cultural critique of research methodologies: a guide to indigenist research and its principles. *Journal of Native American Studies*, 14(2), 109–22.

—— 2006. Indigenist research and aboriginal Australia, in *Indigenous Peoples' Wisdom and Power: Affirming our Knowledge Through Narratives*, edited by J.E. Kunnie and N.I. Goduka. Aldershot: Ashgate, 32–48.

Rosengren, D. 2002. On indigenous identities: Reflections on a debate. *Anthropology Today*, 18(3), 25.

Russell, B. 1946. *History of Western Philosophy*. London: George Allen and Unwin.

Sen, S., Angell, B. and Miles, A. 2000. The Bangladesh Resource Centre for Indigenous Knowledge and its network, in *Indigenous Knowledge Development in Bangladesh: Present and Future*, edited by P. Sillitoe. London: Intermediate Technology Publications, 213–18.

Shiva, V. 1996. *Biopiracy: The Plunder of Nature and Knowledge*. Boston: Southend Press.

Shore, C. and Wright, S. 1999. Audit culture and anthropology: Neo-liberalism in British higher education. *Journal of the Royal Anthropological Institute*, 5, 557–75.

Sillitoe, P. 1979. *Give and Take: Exchange in Wola Society*. Canberra: Australian National University Press, and New York: St. Martin's Press.

—— 1988. *Made in Niugini: Technology in the Highlands of Papua New Guinea*. London: British Museum Publications and [1989] Bathurst: Crawford House Press.

—— 1998. The development of indigenous knowledge: A new applied anthropology. *Current Anthropology*, 39(2), 223–52.

—— 2002. Globalizing indigenous knowledge, in *'Participating in Development': Approaches to Indigenous Knowledge*, edited by P. Sillitoe, A. Bicker, and J. Pottier. London: Routledge (Association of Social Anthropologists' Monograph Series No. 39), 108–38.

—— 2003. *Managing Animals in New Guinea: Preying the Game in the Highlands*. London: Routledge (Environmental Anthropology Series Vol. 7).

—— 2006. The search for relevance: A brief history of applied anthropology. *History and Anthropology*, 17(1), 1–19.

—— 2007. Anthropologists only need apply: Challenges of applied anthropology. *Journal of the Royal Anthropological Institute*, 13(1), 147–65.

—— 2010. *From Land to Mouth: A New Guinea Highland Agricultural 'Economy'*. New Haven: Yale University Press.

—— 2012. From participant-observation to participant-collaboration: Some observations on participatory-cum-collaborative approaches, in *The SAGE Handbook of Social Anthropology*, edited by R. Fardon et al. London: Sage Publications, 183–200.

Sillitoe, P. and Bicker, A. 2004. Introduction: Hunting for theory, gathering ideology, in *Development and Local Knowledge: New Approaches to Issues in Natural Resources Management, Conservation and Agriculture*, edited by A. Bicker, P. Sillitoe and J. Pottier. London: Routledge, 1–18.

Sillitoe, P., Bicker, A. and Pottier, J. 2002. *'Participating In Development': Approaches to Indigenous Knowledge*. London: Routledge (ASA Monograph Series No. 39).

Sillitoe, P., Dixon, P. and Barr, J. 2005. *Indigenous Knowledge Inquiries: A Methodologies Manual for Development*. London: Intermediate Technology Publications, and Dhaka University Press.

Simpson, M. 2001. *Making Representations: Museums in the Post-Colonial Era.* London: Routledge.

Smith, L. Tuhiwai 1999. *Decolonizing Methodologies: Research and Indigenous Peoples.* London: Zed Books.

—— 2005. Building a research agenda for indigenous epistemologies and education. *Anthropology and Education Quarterly*, 36(1), 93–5.

Spivak, Gayatri 1988. Can the subaltern speak? in *Marxism and the Interpretation of Culture*, edited by C. Nelson and L. Grossberg. Chicago: University of Illinois Press, 271–315.

Stanfield, J.H. 1985. The ethnocentric basis of social science knowledge production. *Review of Research in Education*, 12, 387–415.

Stanley, N. (ed.) 2007. *The Future of Indigenous Museums: Perspectives from the Southwest Pacific.* Oxford: Berghahn.

Survival International 2006. Botswana: Bushman case. Source: http://www.survival-international.org/news.php?id=1816, accessed: 10 December 2010.

Swanton, J.R. 1928. Social organization and social usages of the Indians of the Creek Confederacy. *Forty-second Annual Report of the Bureau of American Ethnology to the Secretary of the Smithsonian Institution, 1924–25.* Washington: Government Printing Office, 23–472.

Swisher, K.G. 1998. Why Indian people should be the ones to write about Indian education, in *Natives and Academics: Researching and Writing about American Indians*, edited by D. Mihesuah. Lincoln: University of Nebraska Press, 190–200.

Umilliko Indigenous Higher Education Research Centre n.d. Source: www.newcastle.edu.au/centre/umilliko/indigenousresearchmethodology/index, accessed: 7 June 2012

United Nations 1993. UN General Assembly Resolution 45/163. Passed 21 December 1993, and proclaiming the International Decade of the World's Indigenous People.

Universidad Intercultural Amawtay Wasi 2004. *Learning Wisdom and the Good Way to Live.* Quito, Ecuador: Imprenta Mariscal.

Villegas, M., Neugebauer, S.R. and Venegas, K.R. (eds) 2008. *Indigenous Knowledge and Education: Sites of Struggle, Strength, and Survivance.* Cambridge: Harvard Education Press.

Wangoola, P. 2012. Mpambo Afrikan multiversity, dialogue and building bridges across worldviews, cultures and languages, in *Anthropologists, Indigenous Scholars and the Research Endeavour: Seeking Bridges Towards Mutual Respect*, edited by J. Hendry and L. Fitznor. Abingdon: Routledge, 28–43.

Warren, D.M., Slikkerveer, L.J. and Brokensha, D. (eds) 1995. *The Cultural Dimensions of Development: Indigenous Knowledge Systems.* London: Intermediate Technology Publications.

Wolfe, P. 1994. Nation and miscegenation. Discursive continuity in the post-Mabo era. *Social Analysis*, 36, 93–151.

PART I
Engaging with Indigeneity

Chapter 2

Sharing Anthropological Knowledge, Decolonizing Anthropology: Emberá Indigeneity and Engaged Anthropology

Dimitrios Theodossopoulos

In the early twentieth century, a colonial, and nowadays unpopular, vision of anthropology promoted the documentation of local knowledge for academic audiences. In the early twenty-first century, engaged anthropology is challenging this unidirectional flow of knowledge, encouraging a reverse process of translation from academic language into local (or indigenous) meaningful terms. Making academic knowledge available to indigenous respondents-*cum*-interlocutors often requires such an act of translation: a commitment undertaken by the anthropologist to share or 'communicate back' (in an easily comprehensible form) academic interpretations of ethnography about local/indigenous histories or cultural practices, which may have been, in some cases, partially forgotten, or fallen out of use.

The communication or sharing of anthropological knowledge with indigenous groups – the people who have been previously the subjects of anthropology – can open the way for 'reverse anthropology' (see Sillitoe this volume): a new view of the world that does not strictly comply to Western interpretations. Such an ethos of sharing can make three important contributions to the development of an engaged perspective in anthropology. Firstly, it challenges the divide between engaged and disengaged anthropology, a division that also indirectly reproduces the West and the rest duality. Secondly, it advances the project of decolonizing anthropology by shifting the direction of the translation of knowledge to serve, not the curiosity of academics, but the people studied by academics. Thirdly, sharing knowledge about indigenous culture, history and identity can aid in the articulation of new (often unarticulated) narratives of indigeneity, a process that can potentially empower indigenous cultural representation.

It is widely argued that information collected by anthropologists has helped indigenous groups win struggles over human rights, land entitlement, and legal issues (see among many, Strang 1997; Kirsch 2002; Demian 2003; Hale 2006, 2008; Eriksen 2006; Langton this volume; and in Panama, Howe 2009). Yet, several generations of anthropologists have been ambivalent about making apparent the practical contribution of their research to the greater public, a hesitation that has confined much anthropological knowledge to academia (see MacClancy 1996; Eriksen 2006). An escape from such a dated and exoticized image of anthropology necessitates engagement with 'contemporary realities, in ways meaningful to subjects and readers' (MacClancy 1996: 46), and also includes the possibility, which I highlight in this chapter, that anthropology can interest and be of use to the people studied by anthropology.

To contextualize these broader arguments, I focus on the indigenous identity of an Amerindian group, the Emberá, and their interest in my work and that of other anthropologists. Sharing anthropological understandings about the Emberá with the Emberá has had a transformative effect on my career as it has encouraged me to walk the fine line that separates engaged and academic anthropology. It has also enabled the Emberá in the particular community I study – Parara Puru, Chagres National Park, Panama[1] – to enhance their cultural representation in indigenous tourism. This reciprocal relationship has influenced my academic priorities, and has encouraged me to collect anthropological knowledge about Emberá declining cultural practices (written in academic format, and published in academic venues) to share with them.

Indigeneity, Engaged and Disengaged

The meaning and definition of indigeneity – even the term itself – have been the focus of much debate in anthropology recently, which reflects and further reproduces the polarity between engaged and disengaged anthropology. A preamble to the debate was an article by Beteille (1998) that points to some of the inconsistencies inherent in the notion of indigenous people. Early anthropological terms such as 'tribal' and 'primitive' people – the members of 'simple', 'preliterate societies' – echo the ills of social evolutionism and have been replaced by the alternative term 'indigenous people', which denotes 'priority of settlement' (ibid.: 188). One inconsistency is that in some parts of the world the indigenous people are not in fact the first settlers. The Emberá, for instance, were once confined to Colombia, but they have expanded in the last couple of centuries to occupy lands in Eastern Panama once inhabited by other indigenous groups. In some places, Panamanians of African descent can claim prior settlement to the Amerindian Emberá, although they cannot raise a claim of indigeneity.

---

1 I have conducted 17 months of fieldwork in Parara Puru – spread over seven years from 2005 to 2012 – during which I examined Emberá social change on a variety of topics, such as the use (or not) of the Indigenous attire (Theodossopoulos 2012, 2013a), the Emberá dancing tradition (Theodossopoulos, 2012), the elusive concept of authenticity (Theodossopoulos 2013a), the perception of Indigenous culture (and also the tourists themselves) as 'resources' in tourism (Theodossopoulos 2010), the response of the Emberá to tourist expectations (Theodossopoulos 2011), and the exoticization of indigeneity in the tourism imaginary and the parallel exoticization of the tourists by their Indigenous hosts (Theodossopoulos 2014).

Inconsistencies such as these inspired Kuper (2003) to highlight that the term 'indigenous', apart from its lack of definitional clarity, has similarities with some of the not politically-correct terms it replaced. For him the notion of 'indigenous people' is a modern, more polite adaptation of the term 'primitive'. He lists examples of incorrect use of the term by non-anthropologists that show a romanticized, essentialist, evolutionist conceptualization, and, finally, proposes that anthropologists should abandon the term. His proposition provoked a variety of responses concentrating on the potential implications of Kuper's position for the indigenous peoples' movements or his lack of attention to the history of discrimination experienced by dispossessed indigenous groups (Robins 2003; Place 2003; Ramos 2003; Kenrick and Lewis 2004a, 2004b; Saugestad 2004; Turner 2004; Zips 2006). Some commentators (for example Barnard 2004, 2006) have attempted to reach a compromise between Kuper's purely academic concerns and the importance of indigeneity as a conceptual tool in supporting indigenous rights.

Regarding engaged anthropology, Kenrick and Lewis (2004a, 2004b) criticized Kuper's politically disengaged position for stereotyping the complexity of indigenous realities. They also charged Kuper for not taking into account the historical processes that have shaped the position of some marginal populations, and for underestimating the role and importance of indigenous people's movements. Central to their argument, which they elevate to a criterion for indigeneity, is their experience of dispossession and discrimination at the hands of more powerful groups. Dispossession is also one of the principles of the United Nations Working Group on Indigenous Populations.

Focusing on the internationalization of indigeneity, Merlan (2009) discusses aspects of the United Nations draft declaration on indigenous rights, which considered criteria such as self-determination, settlement and territory, cultural and political rights, re-settlement and occupation. She distinguishes between 'criterial' (based on predetermined, global criteria[2]) and 'relational' (defined contextually according to a relationship with Others[3]) definitions of indigeneity. Saugestad (2001), Kenrick and Lewis (2004), and Barnard (2006) also propose 'relational' understandings of indigenous identities as good alternatives to the essentialism of narrower definitions. As Merlan observes, the discussion (and published literature) on indigeneity has expanded to such an extent[4] that 'there is not just one concept out there but a range involving different histories and positions, that in the interests of international mobilization, are often treated as they were one' (Merlan 2009: 320).

---

2   Inherent to the people defined as indigenous; thus such definitions have a proclivity towards essentialism.

3   An approach towards defining indigeneity that is more flexible and dynamic, as it acknowledges that particular definitions are the result of particular historical processes.

4   The anthropological literature on indigeneity has grown to such an extent that Merlan (2009), in an extensive review, does not refer to seminal articles by Beteille (1998) and Kuper (2003) who published previously in the same journal.

As this short review demonstrates, an engaged perspective towards indigeneity does not only involve acknowledging the political legitimization and internationalization of the term in the context of debates about indigenous rights, but also its contextual specificity and complexity. Local understandings of indigenous identities often have essentialist connotations – as Kuper (1988, 2003) and others observe – and can be profoundly exclusivist, even nationalist. Yet, discarding a term loaded with local meaning contradicts anthropology's fundamental commitment to respect cultural difference. Engaged anthropology, as an orientation that takes local priorities seriously, has no other option than to accommodate indigeneity as defined in local and relational terms, which can potentially include essentialist, nationalist, even racist local views. We can identify, criticize, and contextualize those – often within histories of essentialist, nationalist or racist discrimination effected by more powerful Others – but we cannot ignore them.

Emberá Indigenous Identity: Once a Burden, Today an Asset

The ancestors of the Emberá, known among the non-indigenous Panamanian majority as the *Chocoes* (the people from Chocó, Colombia), and referred to, as other Indian groups, by the denigrating term '*indios*', suffered persecution and discrimination during colonial times. They managed to survive by adopting strategies of migration and dispersion, preferably in relatively remote places in the rainforest (Williams 2005). In the last 40 years, they have formed concentrated communities with schools and a certain degree of political representation (Kane 2004; Herlihy 2003; Velásquez Runk, 2009). Their most visible political achievement has been the establishment of a semi-autonomous reservation in two demarcated districts (the *Comarcas* I and II) in province Darien. Yet, half of the Emberá live outside their semi-autonomous reservation (Colin 2010) and strive to achieve recognition for their lands in Panama.

Following a well-established cultural tradition of migrating along river systems – a response, in most cases, to persecution, over-population, or avoidance of internal conflict – the Emberá have spread from lowland Colombia to Eastern Panama, and as far east as the Chagres National Park close to the Canal, where Parara Puru is located. More recently, some Emberá have entered into tourism, inviting international visitors to their communities to experience presentations of indigenous culture. Indigenous identity has become a valuable resource that enables the Emberá to gain a respectable income without migrating to the city or abandoning their way of life (Theodossopoulos 2007, 2010, 2011). At the same time, tourist interest in indigenous identities has contributed to making Emberá culture more respected at a national level and more visible internationally.

This new cultural visibility is a transformation of status. The admiration of foreign audiences has facilitated a shift in Emberá representation, one that has progressively moved away from the previous stereotype of the '*indio*' (Indian) and closer to the term '*indigenas*' (indigenous), a term denoting a certain degree of acceptance and an acknowledgement of rights. In those Emberá communities that have engaged with tourism – such as Parara Puru – the admiration of foreign visitors, most of whom are individuals from wealthy and powerful nations of the North, has instilled a newly developed sense of indigenous pride. 'The *gringos* want to visit us, the indigenous, not the villages of the latinos', explains one of the leaders of Parara Puru, while his cousin, who was once attracted to the opportunities of the big city, adds: 'I now enjoy dancing and playing Emberá music every day; every day I see again how beautiful my culture is'.

For the Emberá indigeneity is now an asset that can open the doorway to new opportunities. But these advantages have attracted envious comments from critics – for example, urban Panamanians, or some off the beaten-track travellers – who see any aspect of indigenous culture that is made available to tourists as less authentic and real (see Theodossopoulos 2013a). Such criticisms relate to an essentialist view of 'real' indigenous identity as pure and uncontaminated, attained only by isolated tribal peoples untouched by modernity, an exotic view that does not challenge Western civilizing priorities (Ramos 1998: 84).

The tourist quest for the authentic indigene entails a distinction between 'real' and 'unreal' identities, which implies a hidden evolutionist distinction between higher and lower culture. In many respects, the Western idealization of indigenous identities nostalgically conceived as the idealized 'vanishing savage' and lost worlds unaffected by (Western) civilizing processes (Clifford 1986; Rosaldo 1989) sets new dilemmas for the Emberá (see Theodossopoulos 2011, 2013a), who are now expected to satisfy the desires of an international audience wishing to 'consume' Emberá identity. While in the past the Panamanian majority expected them to underplay their cultural difference and become modern citizens of their nation, in the last 20 years the Emberá have realized that the international community expects them to put their indigeneity at the forefront of their representation. In this context, anthropological knowledge can provide valuable inspiration for local narratives about indigenous identity.

Emberá Ideas of Indigeneity

In the Northern Emberá dialect spoken in Panama, Emberá means 'a person' (Mortensen 1999: 1) or 'the people' (Torres de Araúz 1966: 7) or more liberally 'the humans'. It signifies a language and an ethnic group (ideally endogamous). From the Emberá point of view the only other people sharing qualities of Emberá-ness (and nominally acceptable as marriage partners) are the Wounaan, a linguistically and historically related ethnic group (cf. Velásquez Runk 2009). Other people they classify as *kampunia* (that is, no Emberá) (see Kane 1994, Theodossopoulos 2013a). The *kampunia* are further subdivided into Spanish-speaking black people (*kampunia paima*) and Spanish-speaking white or mestizo people (*kampunia toro*), who the Emberá of Eastern Panama also call *colonos* (because they came from

Western Panama to colonize their land). More recently, one hears a new adaptation of the *kampunia* term: *kampunia-gringo,* to describe white North Americans and Europeans.[5]

The Kuna, known in Emberá as '*hurá*', were once traditional enemies, but more recently they have become allies in struggles to secure indigenous rights. The Emberá refer to other Panamanian indigenous groups by the names these groups use for themselves, such as 'Ngäbe', a group in Western Panama whom they Emberá encountered more recently. All Amerindian groups of Panama – including the Kuna, the Ngäbe, and the Wouanan – are recognized by the Emberá as indigenous. Here, the term indigenous is also a racial signifier referring to all people of Amerindian descent: black or white Panamanians (or foreigners) can never be indigenous. Despite my committed efforts to strictly follow Emberá conventions, I have only succeeded in being called 'white Emberá' (*Emberá toro*), but never 'indigenous'.

The Emberá understanding of who is indigenous accords with wider Latin American 'structures of alterity' (Wade 1997: 36–7), which associate indigenousness with 'Indian-ness' and an ancestral connection with the Americas. When they migrate – from Colombia to Panama or from the rainforest to the city – they remain indigenous for as long as they remember their language and they desire to associate with an indigenous identity. Children raised in Panama City who do not speak Emberá are absorbed to the flexible category of *mestizo*, and may become '*latinos*' (Spanish-speaking, non-indigenous people), as my Emberá friends admit with pointed remarks when they criticize relatives or friends who choose to reside in the city.

Overall, the Emberá do not favour the idea of ethnic mixing. Those who marry non-indigenous Panamanians are seen as lost to the Emberá culture and the children of such unions are expected to lose touch with their indigenous identity.[6] Such children are often called *enrazados* (half-breeds), which indicates a racialized view of indigeneity and Emberá-ness. Yet, and to complicate the matter further, cultural factors are also considered important for establishing an Emberá identity. Several offspring of mixed Emberá–non-indigenous unions in Eastern Panama, who have been raised and remained in Emberá communities, are accepted by their communities as full members and a few hold positions of responsibility. Nevertheless, in everyday life, individuals of mixed descent – adults or children, even children or grandchildren of individuals of mixed descent – often become the

---

5   For more detailed information about the uses of the term *gringo* in Panama, see Theodossopoulos 2010b; black North Americans are often described as *gringo-paima* (black *gringos*).

6   The term *campesino* (farmer) – which from the Emberá point of view is synonymous to the terms *latino* or *colono* – is often used with irony to refer to those Emberá who have married latinos or adopted non-Emberá lifestyle. An Emberá would say for example: 'My brother has married a *kampunia* woman and has become a *campesino*; he now lives with the latinos'.

target of jokes that pick up elements of their non-fully-indigenous appearance; for example curly or wavy hair, whiter or darker skin colour.

When non-indigenous Panamanians use the term 'Indian' to refer to the Emberá they perceive it as carrying negative connotations, associated with social inferiority and a less 'civilized' status. From an Emberá point of view, the term 'indigenous' is more politically correct, acknowledging rights and historical connections with the Americas. Unlike 'Indian', which was used in a discriminatory way in the past to 'exclude' or stigmatize inferiority, the term 'indigenous' is associated with the more recent politics of indigenous representation, and the official nomenclature of the nation state, where the term 'indigenous' conveys a certain degree of respect and official acceptance.

An Emberá schoolteacher, who after finishing his degree in Panama City has chosen to work at a local school in Darién, expressed the following views with respect to the term 'Indian' (which are shaped from reading the history of the *kampunia*-'whites', but diverge from the *kampunia* narrative):

> The term Indian that we use for the indigenous did not exist in the Americas. The concept of indigenous did not exist in our language. The term indigenous is a word of the latinos. For 400 years now they call us Indians. This is not correct. In Europe and in the United States people use the term Emberá. Here in Panama they call us *indios*; I want them to know me with my true name, Emberá, my true culture.
>
> When I was in the University, I wrote in my dissertation: "American Emberá culture" (*cultura Americana Emberá*). I told them, the Indians live in another continent, in India. We, the Emberá, live in another continent, the Americas. When they talk about Indians it is like they talk about another planet. The blacks and the whites came from another continent and now they call me *indio*.

I have heard similar complaints from older Emberá who are unhappy with the Panamanian preference to call them Chocoes (the people from Chocó, Colombia). 'It is as if they try to remind us that we came from somewhere else', said an Emberá from Parara Puru, 'but when my grandfather came here there was no border between Colombia and Panama'. An older man expressed the same complaint more directly:

> The world does not know me as an Emberá. Here in Panama they do not know me with the name of my culture. Here in Panama they call us Chocoes. In Emberá culture no one calls anyone Indian: this word is not a word of ours! We are Emberá: the people of this land.

Today the word Emberá is increasingly used in Panama, even by non-indigenous Panamanians. Two separate processes have contributed to this change. Firstly, the formalization of Emberá political structures, which included the appointment of Emberá representatives at community and regional level, where the Emberá, like all other Panamanian groups, are referred to by their self-chosen name. Secondly, the development of indigenous tourism in the last 15 years supported by national advertisement campaigns that present Emberá culture as an asset to Panamanian diversity. As I have mentioned earlier, the new visibility and revaluation of Emberá culture – which is for the Emberá a new positive experience – has inspired a desire among the Emberá of Parara Puru to rearticulate, but also learn more facts about, their cultural identity. It is here that anthropological knowledge can make a contribution.

The Emberá Interest in Others

'Are there any indigenous people in England?' is a question I have been asked many times during fieldwork in Parara Puru. Even, when my interlocutors know the answer – because they asked me the year before – they are keen to compare notes on the human geography of distant countries, or hear me repeating what is for them unusual – almost exotic – information, such as, for example, that houses in England have central heating and that people wear many layers of clothes in the winter. When I remind my interlocutors that I was born in a somewhat warmer country called Greece, I invite the inevitable question: 'Are there any indigenous people in Greece?'

According to the Emberá vision of indigeneity, with its racialized and America-centred orientation, it makes no sense to talk about European ethnic groups – such as, for example, the Welsh or the Basque – as 'indigenous'. So I have developed, in time, diplomatic responses, to explain to the Emberá that there are not any people from the Americas – such as those the Emberá recognize as indigenous – who are native to Europe: Europe is another continent, not part of the Americas. The Emberá listen carefully to my long answers and think, and sometimes they comment – while looking at pictures on my computer screen – that certain ethnic groups, such as the Maori, are indeed indigenous, while others, such as many African people are not indigenous, since according to their ethnic classification they are 'black people' (*kampunia-paimá,* 'black Others').

Through their experience of entertaining foreign tourists, the Emberá of Parara Puru are now aware of several nationalities: during the tourism encounter they carefully observe their visitors, catering to their needs or attempting to anticipate their expectations (Theodossopoulos 2011). In time, the Emberá have developed their own stereotypes about the various foreigners they meet, which are based on their astute observations: visitors from the United States spend lots of money buying indigenous artefacts, the Italians are jolly and flirtatious, the Germans have a tendency to break away from their group and get lost. Such observations show that exoticization based on stereotyping is not the sole privilege of the West, and that indigenous actors can exercise a certain degree of discursive agency, characterizing people from economically more powerful countries (Theodossopoulos 2014). Overall, however, the growing familiarity of the Emberá with new categories of people has excited their desire to expand their geographical horizons and learn about new Others. The Emberá often ask me for additional information, their own anthropologist, who professes to know about other cultures; this is, after all, as I have explained to the Emberá in previous conversations, what anthropologists teach.

It is difficult, I admit, to satisfy the Emberá appetite for geographical knowledge. Most questions address fields of experience that are important to them, such as, for example: How are the rivers in England? What type of fish do they sustain? What animals live in the forest? Are people allowed to hunt? What is the climate? But as the discussion progresses, questions expand from the environment to the economy: What are the salaries of people in this or that European country? Do people own their own cars? What are the prices of clothes or food? The most unusual among my replies – which are often seen as amusing from an Emberá point of view – excite questions that can be repeated several times over the course of the same week, because knowledge circulates in the community, and different individuals want to hear a particular description directly from me. Discussions usually start while I am sitting with the Emberá observing tourists from different countries – for example, Germany, Italy or France – about which the Emberá are interested to know all sorts of factual details.

On my annual returns to Parara Puru, my life and work (in England) and my travels (in Panama or abroad) are a topic of particular interest. I am expected to show photographs on my laptop screen of friends and relatives, other indigenous people I have met in faraway countries or in Panama. I receive progressively new and more detailed questions about the country and city where I now live, my university and students, my work as an anthropologist. I am often asked to describe the courses that I teach (in simple terms), how I use Emberá culture as an example to educate students, the interest of European students in learning about indigenous cultures. In response, I am constantly encouraged to tell the world about the Emberá, persuade more people to visit Parara Puru or provide support.

Among my travelling experiences, those that most interest the Emberá are those about other Emberá communities, both those with which they are familiar, but more particularly, distant communities that they have never visited. Every time I return to Parara Puru after visiting Darién, the province in Panama with the most substantial Emberá population, I am expected to offer detailed accounts of the people, the rivers, the cultivations, and more importantly, famous or less famous shamans I have met. The Emberá gather around my laptop screen to see photographs and videos from other parts of the Emberá world, commenting on people and objects, evaluating or debating the information. This kind of sharing information is not just an invaluable lesson for me, but a source of valuable insights for the community, whose members rely on such comparative knowledge – regarding communities that they cannot afford to visit – to enhance their cultural representation in tourism.

Sharing Anthropological Knowledge about Emberá with the Emberá

On an early trip to Parara Puru I took with me in the field Stephanie Kane's (1994) monograph about the Emberá to provide me with an opening to describe what anthropologists do. The Emberá were thrilled to realize that books had been written about their culture and searched the pages for photographs with excitement. The following year I returned with Donald Tayler's (1996) monograph about the Wouanan in Colombia, which contains a good selection of photographs and figures of artefacts. I was asked to translate extracts, so a year later I brought to the community two Emberá monographs in Spanish, by Torres de Araúz (1966) and Reverte Coma (2002). The residents of Parara Puru received, once more, the photographs and sketches of artefacts with excitement, offering their interpretation of adornments and objects, while agreeing or disagreeing with the transliteration of Emberá pronunciation.[7]

Figure 2.1   Emberá woman in Parara Puru reading Reverte Coma's ethnography

---

7   Yet, they paid less attention to the text, despite the fact that most can read Spanish; it was my job, I realized, to provide concise summaries, as I had previously done with books written in English.

Encouraged by this enthusiastic response, I started systematically collecting visual material about the Emberá from ethnographic sources, scanning old photographs in the British library or photocopying pages from rare articles I obtained through interlibrary loan. Every time I consulted a new Emberá academic source, I created a folder of figures and photographs on my laptop, which I then shared with the residents of Parara Puru at the next available opportunity. In the last seven years, this practice of sharing visual information about Emberá culture has developed into an ongoing and fruitful conversation about Emberá history and identity during which my interlocutors share their own memories – and linguistic accuracy – and I am expected to summarize anthropological interpretations in more approachable terms.

This interest of the Emberá in anthropological information can be easily contextualized within the broader context of indigenous tourism. During tourism presentations, the residents of Parara Puru have to articulate a narrative about their culture and its history. They explain the methods of constructing traditional artefacts, the dance and music traditions they also perform, and describe the history of their community and aspects of its political organization and social life. Each group of tourists, small or large, is welcomed by one of the community leaders who offers a speech to them about the distinctiveness of Emberá culture and way of life (Theodossopoulos 2011). Until three years ago, the speech was delivered by elected members of the community – those holding positions in the running of community affairs – but more recently, all men and women, have started sharing this responsibility, and take turns to participate in delivering the speech.

The task of explaining Emberá culture to outsiders has become a common aspect of life in Parara Puru, undertaken in a variety of contexts – often less structured than the cultural presentations – such as while transporting visitors in a dugout canoe, guiding them through the rainforest, or chatting with tourists while waiting for the regular presentations to start. The ambition of many is to become guides to Emberá culture, to 'teach' the tourists. This desire reflects an ambition to move beyond hesitant experimentation with entertaining Others and start educating Others: that is, as the Emberá say, 'work' towards making 'Emberá culture known' to the non-Emberá world. This process does not merely enhance the representation of the Emberá culture, but also encourages and facilitates its enrichment with new cultural elements (Theodossopoulos 2013a).

The interest the Emberá have in ethnographic information represents a desire to improve indigenous cultural representation by introducing into it knowledge about declining cultural practices, historical events, and a comparative awareness of internal variation within Emberá culture. The residents of Parara Puru, like those of most other Emberá communities, cannot afford to travel widely and visit many other Emberá communities. Their knowledge of the past, as received from parents and grandfathers, does not extend beyond three, or in some cases, four generations. In addition, many communities outside the semi-autonomous reservations have experienced, prior to the introduction of tourism and the recent re-evaluation of indigeneity, increasing exposure to non-indigenous lifestyles.

Figure 2.2   Emberá children browsing Tayler's ethnography and my fieldnotes

When, for example, the Emberá of Parara Puru had to polish their knowledge of Emberá dancing to prepare for their first cultural presentations for tourists, they compared the memories of elderly women born locally and those of Emberá women born in faraway communities (now married to Parara Puru men). The dances contain numerous and constantly evolving variations,[8] and knowledge of the details of the dance as practiced by troupes in other Emberá communities is important to achieving acceptable and aesthetically pleasing improvisation. My own anthropological notes and comparisons on this topic, which had benefited from traveling extensively among Emberá communities in Panama, and also my substantial video archive of Emberá dances has proved to be valuable resource for Parara Puru women[9] (Theodossopoulos 2013c).

---

8   For a detailed description of the Emberá styles of dancing, and revalorization of the Emberá dancing tradition through tourism, see Theodossopoulos 2013c.

9   Women from other Emberá communities have also asked repeatedly to see my videos of Emberá dances in Parara Puru.

Experienced men who are articulate speakers have on several occasions surprised me with their questions – such as, for example, 'where do the Emberá come from?' – although they know that I have seen them answering those questions countless times during tourist presentations. My interlocutors, aware that I read 'all those books' about Emberá culture, sought any additional facts. In responding, I have had several opportunities to explain, drawing from my reading of academic sources, the diversity of Emberá languages in Colombia, the competition between Emberá and other indigenous groups, such as the Kuna, and the chronology of various events in Emberá history. Conversations like these end, in most cases, with the Emberá reciprocating by sharing fragments of oral history, such as for example, stories about the wars with the Kuna that they heard from their grandparents.

The desire of the Emberá to familiarize themselves with a more precise chronology of historical events sharply contrasts with their common practice of collapsing time in everyday conversation. In their collective historical consciousness the ancient times is a vaguely defined period that incorporates themes from 100 years ago to pre-Columbian times. A similar generalizing tendency results in a collapse of geographical and ethnic variation. In everyday narratives about the past, recently infused with information from primary school textbooks, the Emberá often stand for all indigenous people of the Americas persecuted by European colonialism: some younger Emberá, for example, argue that 'Columbus killed millions of Emberá!' Such selective generalizations make more persuasive the rhetorical arguments that highlight the undeniable aggressive nature of Western colonialism and the exploitation experienced in earlier centuries. Yet, some Emberá repeatedly invited me to comment on the exact experience of their ethnic group in their painful colonial history.

On other occasions, such invitations involve engaging in amateur archaeology, exploring the material remains of the past. After seeing (on my laptop) photos of pre-Columbian ceramic fragments discovered by Emberá in Darién, the Emberá of Parara Puru showed me similar ceramic fragments that they regularly find in Chagres. In 2011, a small group took me to a little island in the nearby Alajuela Lake covered with fragments of broken pottery. Unable to provide a sophisticated archaeological interpretation, I invited Panamanian archaeologist, Thomas Mendizábal, to visit the site. It transpired that the location was an old Spanish town, Venda de Chagres, built close to the Camino Real, one of the main routes of transport in colonial times (see Mendizábal and Theodossopoulos 2012). The Emberá guided us along the remaining paved parts of the Camino Real and solicited our opinion about the suitability of this route as an attraction for tourists. At the end of our walk, we all sat exhausted in the dugout canoe, and the Emberá seized the opportunity to ask questions that they have asked me several times before, this time of Mendizábal, a specialist of times past.

A week later, I witnessed fragments of this information circulating in Parara Puru. Several members of the community approached me to visit and identify other sites with antiquities in the vicinity. 'Indigenous archaeologies', understood as local (unofficial) sets of knowledge that represent an awareness of the material remains of the past, can offer an alternative perspective that can contribute to the decolonization of archaeology (Hamilakis 2008). Here indigenous actors can take the first step in inspiring academic research, using local knowledge to guide archaeologists, while archaeologist can share their academic knowledge with local populations, a practice that may lead towards the development of an 'engaged archaeology' (Murray 2011).

Between 2009 and 2010 I worked with the Emberá to prepare a collection of photographs illustrating Emberá practices – such as canoe construction and basket weaving – which the tourists cannot experience directly. The Emberá provided a text explaining the activities, which I translated into English and organized appropriately to support the photographic exhibition set on wooden panels. The exhibition was destroyed by a storm in 2011, but the Emberá, seeing my upset when I witnessed that this archive had been lost, consoled me by telling how much they had learnt about their 'own culture' while working with me to set up this project.

Figure 2.3   Photographic exhibition in Parara Puru before its eventual destruction by a storm

On other occasions, information from anthropology books enhanced daily practices in a more direct manner. After showing the Emberá in Parara Puru Astrid Ulloa's (1992) monograph on Emberá body painting in Colombia, which contains an impressive collection of designs, they asked me to laminate some pages for them to use while body painting the tourists. Painting the arms and legs of tourists – or making an indigenous 'tattoo', as the tourist guides advertise the activity – is a standard part of cultural presentations, and provides the Emberá with an opportunity to earn a $2 to $5 tip. The tourists often hesitate in deciding which design to use or ask the Emberá for non-indigenous themes. The laminated pages from Ulloa's book solved some of those problems, as they are now used as a 'menu' for the tourists to choose. 'The pages can help us teach the tourists the designs of our own culture', commented one of the community's leaders subsequently.

Figure 2.4   Emberá children choosing designs from Ulloa's laminated pages

During the search of ethnographic records for visual material about Emberá culture, I discovered childhood photographs dating from the 1960s of some of the older Parara Puru residents. Taken by Panamanian ethnographers, mostly students of Torres de Araúz, who reproduced them in unpublished theses stored in the library of the University of Panama, these photographic illustrations remained undetected for the last 50 years. They included photographs of the now deceased founders and apical ancestors of the Emberá communities in the Chagres region. The act of sharing these photographs with the residents of Parara Puru inspired the telling of detailed narratives about the history of the Emberá in Chagres, which now feature, in their recently expanded version, in tourist cultural presentations.

Other ethnographic information, such as maps and demographic data, can contribute in unexpected ways to indigenous life. One of my kinship diagrams, for example, has been used by a schoolteacher to learn the names of the children in the community and their corresponding families, while another similar diagram has been used by the leadership of Parara Puru for administrative purposes. Whenever in doubt about the exact population of the community, the Emberá will ask me to deliver an up-to-date figure. When high-resolution maps are needed by Emberá leaders to organize political strategies – in the context of their protracted land rights struggle – they know that they can rely on 'their' anthropologist to procure maps from the geographical institute and planning department in the capital. In this, and many other respects, the Emberá value my scholarly precision, confidence in dealing with bureaucracy, and experience with the technologies of the outside world. The initiation of a Facebook page for the community in 2010, and training one of the community's young men to serve as its administrator,[10] is another example. The hope is that the Facebook page will attract an audience to support future petitions over land rights.

Towards a Non-static View of Engagement and Indigeneity

There was a time when I approached anthropological debates about indigeneity with academic detachment, arguing with a tone of political correctness that the term 'indigenous' should not be used only for the disadvantaged, but for all people's identities. I promoted an all-inclusive view of indigeneity, targeting the essentialist connotations of its exclusive use by particular groups. This view was framed in terms of multicultural politics in urban, Western contexts, but was divorced, as I realized after my involvement with the Emberá, from the concrete political realities of less privileged groups in the periphery of economic power. As with Kuper's (2003) critique, my previous politically correct but detached approach underestimated the meaningfulness of indigenous experience (which is

---

10   Anthropology students from the University of Kent, where I teach, are now able to send messages via Facebook to Parara Puru, demonstrating that they are indeed interested to know about the Emberá and their indigenous culture.

often that of dispossession) and the relational qualities of the term indigenous (see Saugestad 2001; Barnard 2006; Kenrick and Lewis 2004a, 2004b).

My relationship with the Emberá taught me to reconsider some of my previously disengaged views and gradually reshaped my priorities. I learned to appreciate local understandings of indigeneity without fear of their essentialist or exclusivist connotations. While indigenous views of indigeneity may feature fixed and closed narratives that exclude many types of non-fully indigenous people, it is important to appreciate that such local and constantly transforming views are flexible enough to accommodate change. The Emberá, who had been stereotyped until recently as *Chocoes* or *indios*, have now been accepted by the national and international community by reference to 'their own' chosen name. The meaning of being an Emberá in the twenty-first century is not the same as that of the late twentieth century. The term 'indigenous', as understood by the Emberá, carries with it the connotations of such an achievement.

The residents of Parara Puru, encouraged by the warm reception of their culture by tourist audiences, have become aware that the meaning of indigeneity – and its consequences for them – is changing. This is why they are so interested in collecting information about their own culture, including anthropological knowledge from older ethnographic accounts, especially those about cultural practices that had been declining during the last part of the twentieth century. They now welcome information from ethnographic accounts of the past, as well as comparative ethnographic information from the present, which provide a measure of the pace of social change in different Emberá communities. Such knowledge informs new, fluid strategies of cultural representation in the twenty-first century, adapted to cater for audiences of national and international tourists who share contradictory (idealizing or stereotyping) expectations (see Theodossopoulos 2011, 2014).

Indigenous knowledge 'never stands still': it incorporates 'Western knowledge and practices' in the course of the globalization process (Sillitoe 1998: 230; 2002: 15). An awareness of the broader relevance and visibility of one's own culture encourages the forging of alliances between indigenous communities and Western supporters of the indigenous cause (Conklin 1997; Ramos 1998; Brysk 2000). The narratives the Emberá use to describe their indigenous identity, develop as they attempt to articulate their cultural representation in terms meaningful to outsiders and enlarge to incorporate new meanings, including, as I have shown in this chapter, information from academic accounts.[11] In this respect, the fluid and

---

11 The Kuna, the neighbors of the Emberá, have used ethnographic accounts to the benefit of their cultural and political representation. Howe (2009) provides us with a detailed account of the long relationship between Kuna leaders and Western ethnographers. This relationship has led to the development of Kuna auto-ethnography. In comparison, the Emberá have received less attention by anthropologists. They now follow the footsteps of the Kuna in developing their own representational narratives.

constant growth of Emberá narratives of identity can be seen as a transformation of previous transformations (Gow 2001: 127), not as a rupture with Emberá identity.

Seen from this point of view, the renewed interest of the Emberá in the history of their culture, as I have argued elsewhere in detail (Theodossopoulos 2012, 2013a), does not deserve to be interpreted as 'invention of tradition', a notion developed by Hobsbawm (1983) to deconstruct nationalist processes of ceremonialization endorsed by the nation state. The notion of invention can offend indigenous sensibilities and identities (see Linnekin 1991; Shalins 1999; Theodossopoulos 2012, 2013a, 2013b) and undermine the fluidity of social change. Some of the Emberá cultural practices that are now flourishing again due to tourism, had been in decline previously, but they have never completely disappeared. They now expand in new directions through practice, creativity and improvisation, extending the meaning of what is or is not indigenous Emberá culture.

These thoughts lead me to situate the practice of sharing anthropological understandings about the Emberá with the Emberá within a fluid and non-static theoretical contextualization of indigeneity. Anthropologists increasingly question the static myth of the neutral observer, which can be as interfering – legitimizing inaction – as collaboration. 'An action approach, as opposed to a purely academic one' involves 'informing *them* about our thoughts' (Sillitoe 2002: 5; my emphasis). Our thoughts, privileged by information gathered by scholarly and triangulated methods, include information about *them*, often the product of generations of anthropological comparison, corroboration and systematic collection of indigenous knowledge. It is unethical to withhold such information from its original authors.

So we have come full circle. The dividing line between engaged and disengaged approaches to anthropology is not as firm as it seems and it is likely to become less clear with the challenges confronting indigenous studies in the twenty-first century. To escape from imperialist nostalgia, as reflected in the rescue mission of colonial anthropology (Rosaldo 1989) – salvaging vanishing identities, for example – we have to accept social change as the prerogative of indigeneity. Detached anthropological critique can open the way for this process, lifting the burden of a static vision of indigenous society, to encourage a reverse process of translation: previously anthropologists translated Otherness for Western curiosity and academic contemplation; nowadays, anthropology has matured enough – through disengaged deconstruction and self-critique – to embark in a reciprocal engagement with Otherness.

Acknowledgements

I would like to thank the Economic and Social Research Council (research grant RES-000–22–3733) for supporting the fieldwork upon which this article is based, and the residents of Parara Puru for participating with me in a reciprocal quest for knowledge about Emberá culture. I would also like to thank Paul Sillitoe for his constructive comments that have inspired the revision of this chapter.

References

Barnard, A. 2004. Indigenous peoples: A response to Justin Kenrick and Jerome Lewis. *Anthropology Today*, 20(5), 19.

—— 2006. Kalahari revisionism, Vienna and the 'indigenous peoples' debate. *Social Anthropology*, 14(1), 1–16.

Béteille, A. 1998. The idea of indigenous people. *Current Anthropology*, 39(2), 187–92.

Brysk, A. 2000. *From Tribal Village to Global Village: Indian Rights and International Relations in Latin America*. Stanford: Stanford University Press.

Clifford, J. 1986. On ethnographic allegory, in *Writing Culture: The Poetics and Politics of Ethnography*, edited by J. Clifford and G.E. Marcus. Berkeley: University of California Press, 98–121.

Colin, F.-L. 2010. '*Nosotros no solamente podemos vivir de cultura*': Identity, nature, and power in the Comarca Emberá of eastern Panama. Ottawa, Carleton University. PhD dissertation.

Conklin, B.A. 1997. Body paint, feathers, and VCRs: Aesthetics and authenticity in Amazonian activism. *American Ethnologist*, 24(4), 711–37.

Demian, M. 2003. Custom in the courtroom, law in the village: Legal transformations in Papua New Guinea. *Journal of the Royal Anthropological Institute*, 9, 97–115.

Eriksen, T.H. 2006. *Engaging Anthropology: The Case for a Public Presence*. Oxford: Berg.

Geertz, C. 1973. *The Interpretation of Cultures: Selected Essays*. London: Fontana.

Gow, P. 2001. *An Amazonian Myth and its History*. Oxford: Oxford University Press.

Hale, C.R. 2006. Activist research v. cultural critique: Indigenous land rights and the contradictions of politically engaged anthropology. *Cultural Anthropology*, 21(1), 96–120.

—— 2008. Introduction, in *Engaging Contradictions: Theory, Politics, and Methods of Activist Scholarship*, edited by C.R. Hale, 1–30. Berkeley, CA: University of California Press.

Hamilakis, Y. 2008. Decolonizing Greek archaeology: Indigenous archaeologies, modernist archaeology and the post-colonial critique, in *A Singular Antiquity: Archaeology and Hellenic Identity in Twentieth-Century Greece*, edited by D. Damaskos and D. Plantzos. Athens, Greece: Benaki Museum, 273–84.

Herlihy, P.H. 2003. Participatory research mapping of indigenous lands in Darien, Panama. *Human Organisation*, 62(4), 315–31.

Hobsbawm, E. 1983. Introduction: Inventing traditions, in *The Invention of Tradition*, edited by E. Hobsbawm and T. Ranger. Cambridge: Cambridge University Press, 1–14.

Howe, J. 2009. *Chiefs, Scribes, and Ethnographers: Kuna Culture from Inside and Out*. Austin: University of Texas Press.

Kane, S.C. 1994. *The Phantom Gringo Boat: Shamanic Discourse and Development in Panama.* Washington: Smithsonian Institution.

Kenrick, J. and Lewis, J. 2004a. Indigenous peoples' rights and the politics of the term 'indigenous'. *Anthropology Today*, 20(2), 4–9.

—— 2004b. Comment on Kuper, 'The return of the native'. *Current Anthropology*, 45(2), 263.

Kirsch, S. 2002. Anthropology and advocacy: A case study of the campaign against the Ok Tedi mine. *Critique of Anthropology*, 22(2), 175–200.

Kuper, A. 1988. *The Invention of Primitive Society: Transformations of an Illusion.* London: Routledge.

—— 2003. The return of the native. *Current Anthropology*, 44(3), 389–402.

Linnekin, J. 1991. Cultural invention and the dilemma of authenticity. *American Anthropologist*, 93(2), 446–9.

MacClancy, J. 1996. Popularizing anthropology, in *Popularizing Anthropology*, edited by J. MacClancy and C. McDonaugh. London: Routledge, 1–57.

Marcus, G.E. 1998. *Ethnography through Thick and Thin.* Princeton: Princeton University Press.

Mendizábal, T. and Theodossopoulos, D. 2012. The Emberá, tourism and indigenous archaeology: 'Rediscovering' the past in eastern Panama. *Memorias*, 18, 88–114.

Merlan, F. 2009. Indigeneity: Global and local. *Current Anthropology*, 50(3), 303–33.

Mortensen, C.A. 1999. *A Reference Grammar of the Northern Emberá Languages.* Arlington: The University of Texas at Arlington.

Murray, T. 2011. Archaeolgists and indigenous people: A maturing relationship? *Annual Review of Anthropology*, 40, 363–78.

Place, E. 2003. Comment on Kuper, 'The return of the native'. *Current Anthropology*, 44(3), 396–7.

Ramos, A.R. 1998. *Indigenism: Ethnic Politics in Brazil.* Madison: University of Wisconsin Press.

—— 2003. Comment on Kuper, 'The return of the native'. *Current Anthropology*, 44(3), 397–8.

—— 1998. *Indigenism: Ethnic Politics in Brazil.* Madison: University of Wisconsin Press.

Reverte Coma, J.M. 2002. *Tormenta en el Darién: Vida de los Indios Chocoes en Panamá.* Madrid: Museo Profesor Reverte Coma.

Robins, S. 2003. Comment on Kuper, 'The return of the native'. *Current Anthropology*, 44(3), 398–9.

Rosaldo, R. 1989. *Culture and Truth: The Remaking of Social Analysis.* London: Routledge.

Sahlins, M. 1999. Two or three things that I know about culture. *The Journal of the Royal Anthropological Institute*, 59(3), 399–421.

Saugestad, S. 2001. Contested images: 'First peoples' or 'marginalised minorities' in Africa? in *Africa's Indigenous Peoples: 'First Peoples' or 'Marginalised*

Minorities?' edited by A. Barnard and J. Kenrick. Edinburgh: Centre of African Studies, 299–322.

—— 2004. Comment on Kuper, 'The return of the native'. *Current Anthropology*, 45(2), 263–4.

Sillitoe, P. 1998. The Development of indigenous knowledge: A new applied anthropology. *Current Anthropology*, 39(2), 223–52.

—— 2002. Participant observation to participatory development, in *Participating in Development: Approaches to Indigenous Knowledge*, edited by P. Sillitoe, A. Bicker and J. Pottier. London: Routledge, 1–23.

Strang, V. 1997. *Uncommon Ground: Cultural Landscapes and Environmental Values.* Oxford: Berg.

Suzman, J. 2003. Comment on Kuper, 'The return of the native'. *Current Anthropology*, 44(3), 399–400.

Tayler, D. 1996. *Embarkations: Ethnography and Shamanism of the Chocó Indians of Colombia*. Oxford: Pitt Rivers Museum.

Theodossopoulos, D. 2007. Encounters with authentic Emberá culture in Panama. *Journeys*, 8(1): 43–65.

—— 2010a. Tourism and indigenous culture as resources: Lessons from Emberá cultural tourism in Panama, in *Tourism, Power and Culture: Anthropological Insights*, edited by D.V.L. Macleod and J.G. Carrier. Bristol: Channel View, 115–33.

—— 2010b. With or without gringos: When Panamanians talk about the United States and its citizens. *Social Analysis*, 54(1), 52–70.

—— 2011. Emberá indigenous tourism and the world of expectations, in *Great Expectations: Imagination and Anticipation in Tourism*, edited by J. Skinner and D. Theodossopoulos. Oxford: Berghahn, 40–60.

—— 2012. Indigenous attire, exoticisation and social change: Dressing and undressing among the Emberá of Panama. *Journal of the Royal Anthropological Institute*, 18(3): 591–612.

—— 2013a. Emberá indigenous tourism and the trap of authenticity: Beyond in-authenticity and invention. *Anthropological Quarterly*, 86(2), 397–426.

—— 2013b. Laying claim to authenticity: Five anthropological dilemmas. *Anthropological Quarterly*, 86 (2), 337–60.

—— 2013c. Dance, visibility and representational self-awareness in an Emberá community in Panama, in *Knowledge, Transformation and Identity in the Anthropology of Dance*, edited by J. Skinner and H. Neveu Kringelbach. Oxford: Berghahn, 121–40.

—— 2014. Scorn or Idealization? Tourism Imaginaries, Exoticisation and Ambivalence in Emberá Indigenous Tourism, in *Tourism Imaginaries: Through an Anthropological Lens*, edited by Nelson Graburn and Noel Salazar, 57–79. Oxford: Berghahn Books.

Torres de Araúz, R. 1966. *La cultura Chocó: estudio ethnológico e historico*. Panama: Centro de Investigaciones Antropológicas, University of Panama.

Turner, T. 2004. Comment on Kuper, 'The return of the native'. *Current Anthropology*, 45(2): 264–5.

Ulloa, A. 1992. *Kipará: dibujo y pintura: dos formas Emberá de representar el mundo.* Bogotá: Universidad Nacional de Colombia.

Velásquez Runk, J. 2009. Social and river networks for the trees: Wounaan's riverine rhizomic cosmos and arboreal conservation. *American Anthropologist*, 111(4), 456–67.

Wade, P. 1997. *Race and Ethnicity in Latin America.* London: Pluto Press.

Williams, C.A. 2005. *Between Resistance and Adaptation: Indigenous Peoples and the Colonisation of the Choco 1510–1753.* Liverpool: Liverpool University Press.

Zips, W. 2006. Discussion: The concept of indigeneity. *Social Anthropology*, 14(1), 27–9.

Chapter 3

Fostering the Rapprochement of Anthropology and Indigenous Studies: The Encounter of an Italian Anthropologist with Kaupapa Māori Research

Domenica Gisella Calabrò

In this chapter I share my experience as an anthropologist working with Māori academics and encountering Māori approaches to research. During my doctoral studies I investigated the indigenization of rugby in New Zealand and the socio-cultural and political impact of this phenomenon on the definitions of Māori identity (Calabrò 2011). I was enrolled in a PhD programme at the University of Messina (Italy), but moved to New Zealand for a year,[1] where I was hosted and supported by the School of Māori Studies at Victoria University of Wellington. Māori Studies used to be a branch of anthropology and acquired autonomy in 1978, when Sir Mead started the programme of Māori Studies at Victoria University (cf. Mead 1997), directly defying anthropology's monopoly over research methods and indigenous issues.

This positioning enabled me to connect to the Māori academic community, familiarize with their approach to knowledge, known as Kaupapa Māori research, and thus negotiate my space as a researcher. It ended up being a parallel fieldwork. While investigating about Māori and rugby, I was also trying to understand the Māori approach to research and reflecting on my unintentional internalization of colonial attitudes. And I puzzled how and where to locate myself in the Māori academic community as a non-Māori, a non-New Zealander and an anthropologist.

Multiple nebulous zones emerged from this experience. It does not suggest the banning of anthropology, nor does it suppose outsiders necessarily observing and analysing indigenous issues. Conflict and frictions centre, first and foremost, around political and historical issues favouring Western epistemes – a term that Kuokkanen (2007) borrowed from Foucault to emphasize the indissolubility of knowledge and worldview which seems to go unnoticed when we use the term epistemology – over indigenous ones to the extent that many Westerners do not

---

1 I started my PhD in 2007. I was in New Zealand in 2008 (February–December) coinciding with the New Zealand academic year. I went back in the period February–March 2009, in order to observe some Māori rugby events, and shortly visited in July 2010.

recognize that scientific alternatives exist. Negotiations are possible, but they are endless, depending on contingencies, personalities and awareness of the issues.

Genesis of the Journey

My introduction to *Cultural Anthropology* occurred in 2002 in Italy when I was studying Foreign Languages and Literature. It was an optional subject, the exact content of which was unclear to many students. By and large, anthropology represents a niche subject in Italian academia. Like some forefathers of anthropology, I stepped into this world accidentally and indirectly, detouring from my initial studies. What captivated me was that anthropology addressed my intellectual curiosity about cultural variety, and diversity of social issues, while promoting – it seemed to me – respect and mutual understanding of people as cultural subjects.

I learnt about its history, its past faults and its present tensions. The classes also focussed on local relations with anthropological science, when the Italians were 'observed' before being 'observers'. Here anthropological research did not occur within the colonial frame, but I vividly recall images of humble southern Italian peasants – reminding me of my grandparents and the Calabrian reality where I grew up – being described as superstitious and primitive by early observers. My familiarity with Southern Italy enhanced my awareness of the cultural discrimination which pervaded early analyses, and reinforced my empathy towards the 'observed', whose dignity was usually compromised.

The Italian relationship with anthropology is recent, but complex. Local changing attitudes emerged as early as in the 1950s, involving the humanizing approach of Ernesto De Martino. Prior to the post-modern debate over the validity of anthropology's analytical categories and the elaboration of an engaged anthropology, De Martino, who focussed on magic and religion amongst the marginalized inhabitants of southern Italy and the islands (see, for example, De Martino 2005 [1961]), promoted a 'critical reflection on the very categories of our analysis and recognition that these categories derive from our own ethnocentric values' (Saunders 1993: 878), hence, the denomination of his stance as 'critical ethnocentrism' (ibid.). Claiming that all values were historically determined, De Martino engaged in historicizing the lives of southern Italians (1941, 1949); influenced by Gramsci, he disclosed the power differentials that permeated cultural observation (1951). Finally, reflecting on the moral and intellectual implications of research, he suggested that 'one needs to find the way to an ordinary human relation, and situate in the right point where one can be with them in the *same* history' (1999 [1995]: 62).

The legacy of De Martino equally impacted on my predisposition towards the discipline. In this case, it suggested that anthropology had the potential to learn from its mistakes. Nevertheless, being situated on the anthropological periphery, and with my limited anthropological education, I was somehow less aware of the burdensome legacy of anthropology's relationship with colonialism (cf. Asad 1973;

Lewis 1973; Clifford 2004), and the acrimony of some 'classical' anthropological debates (cf. Trask 2001; Obeyesekere 1997). Such degree of naivety played a role in my decision to work with Māori.

If the forces that normally and silently inform one's election of the site had prevailed – such as career outcomes or the topicality of a site, I would have perhaps not dared to propose a doctoral project working in the Māori community. The choice was peculiar in Italian academia. Furthermore, Māori culture scarcely features today in the anthropological landscape. To some extent, the same reason that it fits the archetype of fieldwork (distant, native, tribal) makes it controversial. Also, the emergence of Māori Studies paralleled by indigenous political and social activism has resulted in the alienation of Western researchers from New Zealand's indigenous world (cf. Hanson 1989; Webster 1998).

In New Zealand an anthropologist actually lamented to me that post-graduate students are nowadays unwilling to carry out projects with Māori because they fear them. As I learnt, Māori anthropology students may experience distress too when dealing with Māori Studies. Personally, I have never viewed myself as an intrepid anthropologist. When I sent out my research proposal to the head of Māori Studies, I feared it would be discarded because of my non-Māoriness. At some stage of the fieldwork, I felt uncomfortable. Yet, I was never scared enough not to consider working with Māori or to abandon the field.

I believe the main difference between New Zealanders and I rested in our being situated in different historical and anthropological contexts. I was not familiar with feelings such as guilt for my ancestors' deeds, or fear to discover the unbalance that hides behind a flaunted bicultural credo. I was not affected by the socio-political dynamics that characterize the country and I was not influenced by the misrepresentation of the indigenous community in the dominant discourse, which, I argue, plays an important role in informing non-Māori perceptions of Māori. If anything, I was influenced by the international mis/representation of Māori within rugby, which instilled fascination and curiosity rather than hostility. From this point of view, being an outsider did play a positive role. It allowed my imagination to move freely and shape a collaboration with Māori.

There was also something accidental in my decision to consider a Māori-related topic, albeit it might sound cliché (cf. Gupta and Ferguson 1997: 11). When I started attending anthropology classes, I was already interested in learning about Māori culture. My then partner was from New Zealand. Even though they were not indigenous, he and his family had offered me Māori bone carving and *pounamu* (greenstone) pendants as a gift. That expression of New Zealand's indigenous culture had somehow become part of me. As such, it had been piquing my curiosity. Furthermore, a three-month stay in New Zealand was already planned. Thus, I passionately turned a sojourn motivated by personal reasons into an anthropological enquiry on aspects of Māori culture.[2] Following my studies,

---

2   That first approach to indigenous culture resulted in my four-year degree dissertation, where I focussed on the *marae* and its rituals.

I moved to New Zealand for a year's work experience and learned more about New Zealand ethnic relations and indigenous culture. By the time I decided to undertake doctoral studies in cultural anthropology, I had become quite familiar with New Zealand, its history and social dynamics.

My approach to anthropology and fortuitousness allowed me to picture a research with Māori, yet I believed that I would need the support of the Māori academic community. Both the indigenous mistrust I experienced when I first started to show an interest in Māori culture and the understanding of the historical context I had gained from my stays in New Zealand solicited such a solution. Accordingly, I sent the project's outline to the head of a school of Māori Studies, asking for approval and hospitality. His assent proved that Māori academics do not a priori exclude non-Māori participation to Māori research. As a matter of fact, at that time the school at Victoria was hosting other non-Māori researchers. However, being welcomed to Māori Studies was like being admitted to pass a test rather than passing the test. I was given a chance; the chance to be exposed to Māori research and to experience what is actually about, and to prove that I was worth trusting as a researcher.

The Intelligibility of Epistemic Plurality

When I arrived in Māori Studies, some Māori students recommended I read Linda Tuhiwai Smith's (1999) book about decolonizing methodologies and the formalization of a Māori approach to research known as Kaupapa Māori research. The Māori term '*Kaupapa*' (agenda, philosophy, programme) situates research in a Māori context, integrating their worldview, practices, needs and goals. The Māori understanding of the world, *mātauranga*, rests on distinctive semantic and ontological orders, but early non-Māori observers dismissed it as puerile and irrational (cf. Polack 1840; Thomson 1859), or only valued it as figurative representation of the world – through allegories and personification (Best 1921). In this sense, Kaupapa Māori research has restored the dignity of Māori knowledge.

While students and researchers in Māori Studies asserted an indigenous path to knowledge, I actually sensed that the atmosphere in the Māori academic space was quite dissimilar to what I had theretofore experienced. Everyone viewed the Māori Studies' community (students, teachers and administrators) as a family, upholding values such as reciprocity, sharing, humbleness and participation. Importance was given to collective decision-making. The community aspired to *kotahitanga* (unity) in spite of frictions and dissimilarities.[3] As a researcher, everyone felt responsible towards the Māori community at large as well as the people involved in their work. Everyone showed strong critical spirit in an ironic way, making

---

3   For example, some were more inclusive or more hostile towards anthropology and non-Māori researchers than others, but the engagement to decolonization was a common denominator.

engagement and determination to achieve intellectual enfranchisement palpable. There was the presence of *mana* (spiritual prestige/authority) with the *marae* (Māori ceremonial centre) acting as a 'physical metaphor for intellectual domain' (Hokowhitu 2010a: 13). There was respect for the ancestors symbolized by the carved meeting-house, and emphasis on hospitality (defined by Māori values and procedures). Conferences were modelled on the Māori *hui* (meeting), thus revolving around the *marae* and valuing hospitality.

First of all, the Māori Studies community has emerged as a metaphorical *whānau*. The *whānau* is the basic kin-structure of society with a set of defining values and procedures (Metge 1990, 1995). All the more often groups based on common interests rather than descent or kin 'use the word whaanau as symbol and charter, expressing the quality of their commitment to each other and continually reminding themselves of the tikanga (practices and values) which they associate with the whaanau and to which they aspire' (Metge 1990: 74). People in Māori Studies explicitly refer to their department as *whānau*. The same Māori language points to the notion of *whānau* as paradigm of Māori social relations. The concept of social relationship is expressed by the term '*whanaungatanga*'. Many research participants used this term to define the camaraderie of Māori rugby. It is a relationship based on shared experiences and working together, and informed by reciprocity, where people develop a sense of belonging. To refer to the process of establishing such a relationship, Māori use the term '*whakawhanaungatanga*', where *whaka* means 'to cause something to happen, cause to be'. This same term is also used to translate 'relating well'.

The *whānau*-like relationships within Māori Studies provided insights on the Māori-defined research I was reading about in the books. Firstly, the relation between researcher and informant should be a replica of those relationships. Secondly, the responsibility of the researcher towards the community upheld by Kaupapa Māori research is not solely relevant to the political context; it also falls within Māori social dynamics, where the relationship with the community contributes to define the individual. Finally, the values which ideally inform the *whānau* underpin the ethical guidelines of Kaupapa Māori research as identified by Linda Smith (1999: 120): *aroha ki te tangata* (a respect for people); *kanohi kitea* (the seen face, that is commit yourself, participate); *titiro, whakarongo ... korero* (look, listen ... speak); *manaaki ki te tangata* (share and host people, be generous); *kia tupato* (be cautious); *kaua e takahia te mana o te tangata* (do not trample over the *mana* of people); *kaua e mahaki* (don't flaunt your knowledge). As a matter of fact, formulations of Kaupapa Māori research have explicitly viewed the notion of *whānau* as a pivotal element in organizing research (Nepe 1990; Smith 1999; Bishop 1998). *Whānau*-like dynamics make Māori Studies a safe academic environment to Māori, where they do not have to justify who they are and what they do. I argue that this aspect is paramount in legitimizing the existence of Māori Studies. I also believe that detractors of indigenous studies usually miss this aspect, for the simple reason that they have not experienced or managed to empathize with the cultural estrangement and frustration minority groups tend to experience within mainstream contexts (cf. Sandri's chapter).

To continue, Māori research has emerged as an integrated whole, for 'Māori's science system is in a form that cannot be separated from the family system, religion system and political system' (Ito 2004: 20). The Japanese anthropologist Ito (2004) observed Māori Studies as a discipline against Luhmann's social system theory, which describes modern society as being functionally differentiated in many systems, which are autonomous, self-referential and closed, and use binary codes (for example, true/false; legal/illegal). Upholding an all-encompassing approach to science, Māori Studies are viewed as an example of 'dedifferentiation' of systems. Thus, holism emerges as the defining trait of Māori Studies against non-indigenous disciplines. Because of the non-alignment of indigenous studies with the logic of Western disciplines, their scientific integrity is often questioned.

The polarities of Western/dominant groups and Indigenous groups do not circumscribe differences in knowledge, academic institution, and power relations. As a student initiated in anthropology in an Italian institution, I recognize the presence of centre and peripheries within my discipline (Gupta and Ferguson 1997: 25–9;) as well as the impact of my national scientific field 'with its traditions, habits of thought, problematics, shared commonplaces, and so on' (Bourdieu 2003: 283), which caused me to experience a subtle form of estrangement in the mainstream institution where Māori Studies is situated too. Still, despite my academic origins and the awareness of cultural difference which pertains to anthropology, I had not appropriately anticipated the 'cultural shock' at Māori Studies, which shows the extent to which we fail to realize that the academic environment is itself culturally informed, and take many of our procedures and values for granted.

However, I also realized that our tendency to give our research categories and academic practices for granted paradoxically confirms the very cultural dimension of academia. As illustrated by Kuokkanen, episteme '[…] is a mode of social reality, a reality that is taken-for-granted ground whose unwritten rules are learned (or as Foucault would say, "written" in the social order) through the processes of socialization into a particular culture' (2007: 53). From this point of view, I argue that the Māori-enhanced awareness of their episteme is to be ascribed to the historical events that have not allowed them to take it for granted. Māori have had to objectify their episteme, in order to preserve it and safeguard it. Ultimately, this comes down again to power differentials.

Historicizing Māori Studies' Bitterness towards Anthropology

Being entrenched in the Māori worldview, the authority of Māori Studies and *kaupapa* Māori research transcends the here and now. Yet, we have to consider political motives, which are the same as the indigenous movement, and inform the same definition of 'indigenous' as illustrated by Sillitoe's chapter. Māori Studies, like indigenous studies generally, oppose subjugation within the field of knowledge challenging Western disciplinary control (cf. Walker 1996; Smith 1999; Mahuika 2008). Unlike other disciplines, anthropology has esteemed non-Western knowledge for generations, even though mostly framed in Western categories. Similarly, it has tried to dialogue with indigenous communities for a long time. Anthropologists who worked during the colonial era were not usually part of the colonial machine; nor did they necessarily endorse colonialist ideas, so much so that we also find early attempts of self-reflexivity; moreover, the colonial officers did not appreciate anthropologists either.[4] Nevertheless, due to the historical context and the way ethnographies have been used by colonial and postcolonial forces, amongst indigenous people anthropology has gained a reputation as the main academic locus of colonial domination,[5] and tends to be still perceived as 'the study of primitive societies' as it used to be conceived in its origins (cf. Evans-Pritchard 1951; Nadel 1953).

I have accordingly been told or suggested that my research was unethical. Or obsolete. I have been 'challenged' by being asked how I would feel if someone started observing and commenting upon the people I identified with and our culture, being presumed that it would not normally be the case. I responded by admitting I would be initially suspicious too. I shared my community's relation to anthropology. I illustrated the definition of anthropology I embraced. I stressed that it was not my intention to get into a competition between internal and external gaze and to replace the Māori viewpoint. I described my standpoint as complementary to the insider one, provided that no-one is ever totally outsider or insider,[6] and that one's specific experiences impact on research as much as the (not) belonging to the culture investigated. On one occasion I challenged back, arguing that it would have been really interesting to see my culture and the West in general through a Māori lens. How would they perceive facts, values and practices? How would they approach research? In that case, it was pointed out that Māori cannot afford the time to understand other cultures, for they have to daily strive to legitimate and preserve their own, and have to urgently face issues deeply affecting their community well-being.

The conclusion is that to some extent I was able to defend anthropology and to go to the other side of world in order to work with Māori, because I was privileged enough not to have to constantly worry about defending my own culture and

---

4 See Wendy James (1973) and her analysis of Malinowski's writings about the relation between anthropology and colonialism. See also Sillitoe (2006) and his review of the application of anthropology during British colonialism, and Bourdieu (1985) and his observation of the relationship between anthropologists and colonial officers in the French colony of New Caledonia.

5 See Ottenberg (1990: 151) who remembered going to do fieldwork in Nigeria driven by a genuine interest and ideals of recognition and seeing his fieldnotes labelled as colonialist as soon as Nigeria gained independence in 1960.

6 The scientific approach to facts and events create a first element of disconnection from the community we work with (cf. Narayan 1993: 679–80).

identity. In indigenous contexts many grievances remain feeding resentment. Anthropologists' work – decontextualized, misinterpreted and manipulated – may still be misused by media and dominant institutions.[7] Furthermore, the balance of power still favouring *Pākehā* (New Zealanders of European descent) over Māori, the latter can exercise *rangatiratanga* (self-determination) in a meaningful way in Māori contexts only (Metge 1995: 311). Basically, when Māori and I talked about anthropology, we would often talk past each other, our perceptions of anthropology being related to distinct pasts and distinct presents.

It is also interesting noting that the Māori view of culture does not encourage anthropological observation either. All expressions which can be labelled as culture (symbols, rituals, language, arts, knowledge about the world) are classified as *taonga*, a term commonly translated as treasures, precious possessions. They are therefore expected to be treated with respect, preserved and transmitted to the inheritors, and are collectively held and managed by *iwi* (tribe), *hapū* (subtribe), and *whānau*. The consequence is that 'Māori who have internalized this understanding about the nature of knowledge [...] refuse to generalize about Mātauranga Māori and speak only of what they learnt and know as members of their own iwi, hapū and whānau' (Metge 1995: 310). This aspect emerged when interviewing, inasmuch as participants would frequently clarify that they could not speak on behalf of all Māori, or that they could only refer to the experience of their own tribal groups. This makes us reflect upon the fact that the mere act of generalizing upon cultures, talking about and 'handling' cultures different from our own is situated in the Western logic. When we question our research possibilities with Māori, we should maybe 'worry' about this aspect rather than the Māori current opposition to the discipline.

In Māori contexts the critical approach to anthropology as a form of political domination is, however, long-standing. The correspondence between Te Rangi Hīroa (or Sir Peter Buck), the first Māori anthropologist, and the politician and lawyer Apirana Ngata, who played a pivotal role in the resurgence of Māori pride and the improvement of their socio-cultural condition during the first half of the last century, provides an honest and acute observation of anthropology at the time, which remains topical (Sorrenson, vols 1–3, 1986). In 1928 Ngata expressed his distress at the way Anglo-American anthropologists represented Māori and other native peoples negatively (Sorrenson, vol. 1, 1986: 91–2). In response to that comment, Buck advocated the creation of an indigenous approach to cultural analysis (Sorrenson, vol. 1, 1986: 121–2). We also see Buck considering and opposing the common idea that Māori lacked abstract thinking (Sorrenson, vol. 1, 1986: 236; cf. Metge 1976: 72–3), lamenting how Pākehā anthropologists normally glossed over Māori corrections and opinions, whenever they tried to make their voice heard (Sorrenson, vol. 2, 1986: 172), and revealing the Pākehā

---

7   A famous example is the article 'US expert states that Māori culture is an invention' published in 1990 by the New Zealand newspaper *The Dominion Post*, based on Hanson's article (1989), which caused a harsh socio-political debate.

economic and political interests that lay behind the mission to 'civilize' the Māori and the use of anthropology to fulfil those goals (Sorrenson, vol. 2, 1986: 209–10). We even see him reflecting about the possibilities and limits of his view as insider in Māori culture (Sorrenson, vol. 1, 1986: 123–4).

Buck and Ngata questioned the authority of Western anthropologists and their methods and goals, but believed that the anthropological approach to culture could be used to their own benefit and in their own way. At the turn of the twentieth century, following the land wars, Māori had decreased to 15 per cent of the population and lost most of their lands. Morals were collapsing and many Māori were dismissing their practices and language, assimilation being seen as the only option available. Given this scenario, 'Ngata and Te Rangihoa saw Anthropology as a tool for cultural recovery and for expressing and maintaining a deeply-held sense of identity and cultural being' (Kahotea 2006: 6). As a matter of fact, Buck wrote several ethnographic monographs on Māori culture, documenting elements of Māori culture that were disappearing or had disappeared.

At first, I believed Buck and Ngata's stance could legitimize and ease a possible collaboration between anthropology and Māori research/ers today. The Māori anthropologist Kahotea (2006) also referred to Buck's work and approach to anthropology to justify his commitment to the anthropological research. Nonetheless, the scenario I have sketched in the previous paragraph suggests that at their time there was no alternative but to promote change within anthropology. Consequently, the use of research to benefit the community, the reflection on an indigenous approach to research, and the effort to challenge Western determination within research could be rather viewed as a primeval attempt of scientific self-determination and thus an early form of Māori Studies.[8]

In any case, following the Second World War, the socio-political and economic context changed radically and certainly left little space for similar even-handed voices,[9] and Māori definitely got to view anthropology as the stronghold of the

---

8   On the other hand, the Māori leaders were able to play a preeminent role in academy and politics because they participated to the new colonial society. In some cases, Buck would himself adopt stances that mirrored the interiorization of elements of evolutionism and cultural discrimination. For instance, he claimed the dignity and efficiency of Māori culture by referring to their alleged Aryan origins (cf. Hanson 1989). Like Westerners looked down on Māori, he would look down on Africans. These contradictions could be understood viewing Buck as a personality in-between two dissociated worlds, as his double name might suggest. In this sense, not only was the emancipation of Māori research from anthropology unfeasible, imagination could not conceive spaces of socio-cultural inquiry on Māori other than anthropology.

9   Ngata passed away in 1950. No leader able to stir Māori pride and energy emerged after his death. The curriculum of the Māori boarding schools had been changed, in order to orientate Māori towards manual and agricultural labour. New Zealand adopted an industrial economy. This economical change originated the phenomenon of Māori urbanization, which deeply affected Māori socio-cultural realities. New Zealand also ceased to be a colony and the government tried to accelerate assimilation. The Māori language had nearly been lost

Pākehā definition of Māori.[10] The Māori upsurge at the national political level which is not solely about recognition, but is also an expression of indigenous conscientization (Smith, G. 2003), eventually enabled them to emancipate from anthropology. Seeking to elaborate an autochthonous approach cognisant of the Western notion of research, it is arguable that Māori Studies represent the intellectual pursuit of Buck's and Ngata's ideals today. This would corroborate the argument that anthropology is outdated, having been replaced by indigenous studies. Yet, Māori Studies go beyond the way of the scope and interests of anthropology, and the latter discipline is not merely about understanding indigenous cultures. Rather than substituting anthropology, I argue that Māori Studies have gained their own disciplinary dignity and space within the academia.

Facing Politics to Bridge Disciplinary Distance

Since the controversy between anthropology and Māori research stems first and foremost from the historical context, to negotiate a space as an anthropologist I had to encompass the political dimension of Māori Studies, its implications and responsibilities. The fact that I was doing Māori-related research and situated myself in a Māori academic context had political and ethical implications, which fully emerged in the range of reactions and qualms my positioning raised. I do not necessarily refer to Māori reactions. I include the response of anthropologists or students who stay away from Māori topics because of the political reality, and, above all, the reactions of non-Māori New Zealanders, particularly Pākehā.

Non-Māori were startled as much as Māori were by my interest in indigenous elements and issues. A few of them – usually people who were knowledgeable about indigenous culture – applauded my interest and approach. In most cases, I would perceive circumspection and malaise. Some seemed to be experiencing a feeling of inadequacy vis-à-vis the fact that I possessed a better knowledge about Māori than they did. Others seemed to automatically interpret my attempts to learn the Māori language and the Māori ways as my siding with the indigenous people in an alleged battle between Māori and Pākehā. Sometimes I was 'reminded' how Māori culture is unauthentic and merely political, or told that Māori had pushed beyond limits the 'story of the Treaty of Waitangi and the land'.[11] More frequently,

---

and the whole context was discouraging. As a few Māori told me, from the 1950s to the 1970s many Māori downplayed their indigenous identity and neglected Māori practices because 'it was not cool to be Māori'.

10   I must nevertheless mention the contribution of Winiata Maharaia. He gained a PhD in Anthropology in 1954 and upheld Buck and Ngata's ideals and views about the discipline, but prematurely died in 1960.

11   On 6 February 1840 the representative of the Crown and various tribal chiefs met at Waitangi in the North Island to sign a treaty that established British sovereignty in New Zealand, recognized the Māori ownership of their lands and other properties and granted

I was gently made to understand that I could not really appreciate how things worked because I was not a New Zealander.

Consequently, one cannot ignore the politics without de-historicizing Māori hostility towards anthropologists and, more generally Western observers, which, as noted, is the legacy of colonialism. I do not believe this is a reason to put a halt to anthropologically-informed research. When Smith identified four formulas of cultural sensitivity that could accommodate for Western presence, she actually mentioned the 'strategy of avoidance' (1999: 176). It might be argued that if the researchers are not prone to engagement, dialogue and understanding of indigenous needs and practices, avoidance might be preferable. However, it will eventually benefit no-one, as claimed by Smith (ibid.). Avoidance contributes to create and/ or solidify barriers. It also amounts to acceptance of 'epistemological racism (i.e. what is considered legitimate epistemology in the academy) as well as of sheer indifference and ignorance of the sort that takes Western epistemes for granted as the only valid point of departure' (Kuokkanen 2007: 67).[12]

Avoidance is not necessarily indicative of cultural sensitivity either, insofar as fear seems more commonly to dictate it as well as a desire to spare oneself the trouble of complex negotiations, and unwillingness to question one's methodology and values to experiment with diverse approaches (cf. Reilly 1996: 404). From this point of view, avoidance suggests compliance with the dominant view of indigenous people and issues as problems. In New Zealand Māori are perceived as a problem (Walker 1996) and so is the Māori–Pākehā relationship (Maaka 2008), leading Smith to argue that 'problematizing the Indigenous is a Western obsession' (1999: 91).

During my first difficulties and discomfort in the field, I fell victim to this trite discourse, by viewing Māori reluctance towards Western observers and/or anthropologists as a problem. Realizing that I had been conforming with, and thus perpetuating, the dominant discourse had a significant impact on my field-approach. Accordingly, I started to view the situation as one where historical circumstances inflamed suspicion of the researcher, and to focus on why things had been easier in the past rather than the reasons why 'it is hard to work with Māori today'. Indeed, the issue is not so much that 'collaboration' used to be easier because of power imbalance, but about anthropology having been spoiled

---

Māori the rights and duties of British citizens. Māori view land confiscation as a breach of the promises made in the treaty. A permanent commission of inquiry, known as the Waitangi Tribunal, was set up in 1975. The tribunal makes recommendations on claims regarding breaches of the treaty brought by Māori and aiming to compensation or land restitution.

12   My impression was that ignorance and indifference were ingrained into the majority of non-Māori opinions of/attitudes towards Māori. In New Zealand these conditions are discreetly created by mainstream institutions. Ignorance is thus sanctioned (cf. Spivak 1990) and indifference is institutionalized (cf. Herzfeld 1992).

by decades of (apparent) collaboration,[13] which makes it harder to accept and cope with obstructions and enmity.

Thus, rather than discontinuing anthropological research about indigenous-related issues, one should look for or try to build bridges towards collaboration, mutual respect and understanding (cf. Hendry 2007, Hendry & Fitznor 2012) and should conduct research bearing in mind the political implications of their work. While I believe that anthropological investigation becomes political exercise, I argue that its political dimension is quite subtle. It is about understanding the current research conditions as the result of history and politics and being aware of the community's vulnerability; it is about being responsible towards the community, more inclusive of their voices and vigilant as to the way data is handled and ideas are articulated. Refusing to face the political issues reiterates the mistake of anthropologists in the colonial era (cf. Asad 1973; Sillitoe 2006). We can be merciless towards our predecessors, overlooking primeval forms of self-reflexivity and historical constraints, but 'those scholars today who are critical of the colonial mentality of that time are just as likely to be blind to the current political realities in which they are entrapped and for which problem is perception of the current world one lives in; the past seems much easier to understand from within the framework of the present' (Ottenberg 1990: 152).

I found it illuminating to visualize the political arena where Māori and non-Māori meet as a myriad of intersecting trajectories and relationships, rather than a unique trajectory of two cultures opposing each other, as elaborated by the Torres Strait Islander scientist Martin Nakata (2007), who conceived this space as a 'cultural interface'. What I view as an obstacle in the acknowledgement of the historical legacy and the way it impacts on the 'cultural interface' is the enduring negative perception and use of the term 'political'. Many Pākehā implicitly blame Māori strongly asserting their Māori-ness as well as anti-colonial ideas and aspirations by labelling them as political. Likewise, the dominant discourse demeans Māori research by identifying it as mere political activism (cf. Sandri's chapter). What bemused me is that the Māori contexts are not exempt from this perception of the term. I noticed it was not uncommon for Māori to identify the position of more intransigent or radical scholars and thinkers as 'political', even though it would often occur in ironic terms. What about the 'non-radical' researchers? Implying that in that case there were no politics involved contradicted the holism as well as the engagement in decolonization of Māori Studies. In New Zealand the mere act of proudly and overtly stating one's indigenous identity corresponds to a political statement. Hence, we need to review critically the derogative interpretations of the term 'political' as it informs indigenous-related academic debates.

---

13   See Yacin, Mammeri and Bourdieu (2003) and their considerations on the illusions of the fieldworker regarding the collaboration of the observed ones in the colonial context.

Investigating Māori and Rugby: Hows and Abouts

Rugby informed my first impressions of Māori, conveying images of a proud and strong people connected to their heritage while integrated into the egalitarian society embodied by the All Blacks, the New Zealand National team. This occurred in autumn 1999, when England was hosting the first professional Rugby World Cup and I was temporarily living in its capital. In that period a huge billboard of Adidas – sponsor of the All Blacks – was hanging in Tottenham Court Road, one of the busiest streets of London, displaying the tattooed face of a Māori warrior as a symbol of the team. Not long after the billboard had captured my eyes, I also watched my first All Blacks match and their pre-match performance of the *haka* (Māori posture dance), sensing immediately the clamor the team engendered in Europe. When events situated me on the crossroads between Europe and New Zealand and I started to actively get interested in Māori realities, rugby seemed to act as the background leitmotiv. Europeans who found out about my connections with New Zealand would regularly ask me questions concerning the *haka* and Māori as rugby players, voicing their rugby-related opinions of Māori and their stereotypes about indigenous people at large. While in New Zealand, I would look at things against the image vehicled through rugby, noticing that reality partly contradicted such an idyll. I realized that not only did rugby act as an interface between Māori and Europe, it also emerged as the main site where Māori and non-Māori would encounter. Also, I accessed a broader vision of Māori culture and gradually sensed that the relationship between Māori and rugby was much more than the simple Māori participation to a Western cultural practice. I therefore came to view rugby as a site that would provide further insights into the socio-cultural and political experience of Māori as colonized subjects as well as an indigenous minority in a postcolonial society.

This is how a non-Māori woman coming from a nation where rugby is a 'tradition under construction', and who had never been a rugby fan or played this game, ended up probing into the Māori relationship with rugby. My personal journey got me to reflect upon the phenomenon and identify it as a valuable research topic. Anthropology supplied the conceptual and theoretical instruments and the vocabulary to analyse it. Māori Studies provided me with guidance, intellectual sponsorship and the chance to actually interact with the Māori community.

However, my object of investigation somehow made my negotiation of a space more difficult. 'Rugby is just a game' is a refrain one can commonly hear in New Zealand. As such, it tends to be viewed as lacking relevant socio-cultural meaning and being separated from politics. A similar topic may be perceived as trivial, or not impellent enough within Māori research, and even with engaged anthropology. The topic as well as my apparent lack of credentials to analyse it (including being a woman) might have at first seemed to confirm the assumption that Māori are supposed to best know what is significant to them. I tried to demonstrate that perspectives originating from distinct experiences can instead enrich the spectrum of indigenous-related knowledge and I deem that the Māori worldview asserting the interdependence of all aspects of reality, it should sanction such complementarity.

The research confirmed that rugby has been the object of a process of indigenization, aiming to fulfil Māori ends of socio-cultural continuity and political recognition. It has allowed Māori to transmit some of their values and practices as a means of safeguarding and re-invigorating indigenous social dynamics undisturbed. Due to the elevated status of rugby in New Zealand society, rugby has also become a privileged site for Māori to participate in mainstream society and a platform to reaffirm their values and define their identity locally, nationally and internationally. Today, it is also seen as a site of social upward mobility. Nevertheless, rugby has also emerged as being still instrumental to the control of Māori cultural agency, representation and self-determination within the wider society. I therefore individuated an ambiguous and dialectic site, which bears testimony to both Pākehā hegemony and Māori resilience and has ambivalent repercussions on the contemporary indigenous lived experiences. I believe this investigation contributes to disclose elements that impinge on Māori self-actualization and integration as well as to highlight Māori socio-cultural distinctiveness and vitality and their aspirations and needs in contemporary society.

During fieldwork I also tried to encompass aspects of Māori research, moving towards a cross-cultural methodology.[14] By situating myself at Māori Studies, I had unawares adopted the *tiaki* or mentorship strategy identified by Graham Smith (Smith, L.T. 1999: 176). It was easy to also endorse the strategy of 'personal development' (ibid.), whereby the non-Western researcher becomes knowledgeable of and participates in Māori culture, insofar as it is a given-for-granted procedure in anthropological research. Since my arrival I clearly immersed myself in the Māori cultural world, which included learning the Māori language attending both university classes and the community-based *Te Ataarangi* classes[15] at the *marae*. Following the *pōwhiri* or welcome ceremony at the *marae*, I started to take part in the *hui* of the university *marae* as part of the *tangata whenua* (hosts) rather than the *manuhiri* (guests). This included welcoming and helping out during the *hui*.

However, my personal development also involved elements that were not comprised in my training as an anthropologist. I moved towards a collective approach by welcoming the ideas, suggestions and expectations of participants and Māori researchers and being available to answer their questions about the research and my own experience. The interview turned into a nice interaction which first and foremost enriched me as a person. The purported construction of a long-term academic relationship with Māori became more central. The question 'How will the research benefit the Māori community?' became less vague and

---

14   Cf. Gonzalez (2010) and Sakamoto (2011) for other examples of non-Māori researchers trying to be inclusive of Kaupapa Māori research.

15   *Te Ataarangi* is the Māori reinterpretation of the silent way language learning method elaborated by the Jewish linguist Caleb Gattegno, where Cuisenaire rods are used to illustrate the language, spoken language is privileged and learning is student-centred.

more relevant and I grew more responsible towards the people involved in my research. I also learnt to be more patient and flexible, in spite of the pressure of institutional deadlines and parameters.

I did not relate to people as a Western researcher in anthropology, but as an individual whose identity was no less complex and multifarious than the Māori ones. I disclosed my *whakapapa* (genealogy) as an Italian and that enabled people to establish a first connection. In situations like gatherings or first contact, I learnt to disclose it through the *pepeha*, a set of verses in Māori language which defines the individual in relation to their land and their social relations. Some people, adults and particularly elders, saw a connection with my Italian identity because of the historical relationship that the soldiers of the 28th Māori Battalion established with the Italian soldiers and families during the Second World War, in the period corresponding to the German invasion of Italy following the Italian surrender to the Allied forces. Others established a connection because of their work interactions with the Italian diaspora community, or because they themselves had Italian ancestry.[16] From this first connection, people would identify further connections based on what they pinpointed as cultural similarities, such as the emphasis placed on hospitality; the importance of family; the attitude to life; the passion for food. In relation to the former aspect, I confess that something as simple as reciprocating hospitality or support by making a tiramisu revealed to be a tool of connection. Regarding my *whakapapa*, I equally often shared my identification as a Southern Italian and its implications. The history of the area I come from is characterized by elements of oppression, discrimination and poverty as well as of internal conflicts. Still today, in Europe and in the rest of Italy Southern Italians can be the object of negative representations and stereotypes, which echo the accounts provided by early observers.

Moving from that I tried to establish connections based on shared interests (such as Māori rugby or Māori research at large) and experiences, according to the principles of *whakawhanaungatanga*, and gradually created a network. I therefore interviewed people that I had got familiar with or people that other Māori – acting like intermediaries – introduced me to. I abandoned the use of a formal email as a way to ask for an interview to privilege face-to-face or at least phone communication. I entered into the spirit of reciprocity by appreciating their availability through a *koha* (gift) that represented something from my culture whenever possible, and by engaging to send results. I learnt to value *mana* rather than status in Western terms, and tried to respect participants' *mana* by naming them and acknowledging them as a source of my knowledge. I decided to write my thesis in English instead

---

16   The *whānau* Sciascia, which numbers more than 2,000 people, revolves around the eponymous ancestor Nicola Sciascia, a man who migrated from Italy – notably from Trani in the region of Apulia – to New Zealand during the second half of the nineteenth century and married a Māori woman, Riria McGregor. Furthermore, in the 70s, a group of Italians went to New Zealand to construct the Rangipo tunnel and quite a few of them eventually married Māori women and stayed in the country. According to statistics, a higher proportion of Italians than of any other New Zealand ethnic group identify their second ethnicity as Māori.

of Italian, so that it could be accessible to Māori, and committed to acknowledge the Māori language as a *taonga* and as part of the Māori lived experiences by respecting the use of the macron and avoiding italics. Finally, I included lengthy sections of the interviews in the thesis, so that the reader could actually encounter many voices with different personalities, experiences and perspectives. My hope was that the research would emerge as a choral work. I equally hoped the research to be a platform where existing voices could articulate themselves rather than a space where I would purely give voice to indigenous ideas and aspirations. All these adjustments may seem like '*minutiae*', but they also allowed me to initiate a conversation which will hopefully be continued, and to develop a warm sense of belonging on the other side of my world.

Conclusions

My experience suggests that working with Māori as an anthropologist and a non-indigenous researcher is achievable, and to some extent desirable, notwithstanding diversified obstacles, which can be met in both anthropology and indigenous studies. In 2012, during a conference on indigenous research, a Māori anthropologist (Muru-Lanning 2012) talked about the way she carried out doctoral research in her own community in compliance with Māori perspectives, practices and values. This part had to be deleted from her thesis chapter on methodology, for it was not considered to be anthropology. Irony of fate, I had just presented a paper about my own experience as a non-indigenous anthropologist working with Māori, including a brief but passionate apology of current anthropology and its possibilities in indigenous research. If anthropology expects researchers to understand and participate in the cultural realities they analyse, why should not this extend to research practices?

As for me, I have sometimes been classified as a case of 'gone native'. As a response to my ideas/ideals, I have been once told that I did not need to do a PhD and should have instead trained as a social worker. More generally, I have suffered the collision between my patience and my desire to be an accountable researcher in the Māori community and a reality made of rigid timetables, scarcity of funding and specific expectations as to the way things need to be done. These elements raise further questions. Partly, they are related to the way we conceive anthropological distance. In my view, it corresponds to a level of emotional detachment from cultural assumptions, which results from and further fosters the development of the academic approach. It is an uneasy process – for our own cultural experiences lose some of their emotive strength and spontaneity – which enables the individual to analyse cultural issues and develop the *potential* to achieve a degree of insider access to other socio-cultural realities. Partly, the questions are related to the awareness or acceptance of our historical obligations as well as of our disciplinary responsibilities, as a science dealing with human beings and the ways we make sense of our lives. Overall, the overview of current issues witnesses fear of losing disciplinary control (cf. Sillitoe's chapter).

Indigenous anthropologists are probably in the most uncomfortable position. Aiming to be viewed as accountable in both contexts, they are perceived as 'not anthropologist enough' or 'not indigenous enough'. The Māori anthropologist Kahotea (2006) wrote about his dilemmas as to how position himself in the field of anthropology as a Māori. He eventually did not mention Kaupapa Māori research and yet later realized that his role as a native informant respecting Māori practices and his being inscribed in the tradition of Māori anthropology automatically aligned his research to the Kaupapa Māori one.

In indigenous contexts, the issue is the hostility towards what appears as Western. Te Punga Sommerville (2011), a researcher in English literature, noticed that because of the constant association of Kaupapa Māori research to concepts of resistance and disenfranchisement, Māori have tended to restrain from doing research within fields – such as her own (and I would add anthropology) – or about topics that seem not to be useful to Māori, or do not involve Māori or are not associated to Māori 'traditional' ways of being. Such union may even lead Māori to disregard literature written by non-Māori. Rather than wondering 'what it [Kaupapa Māori research] enables', she thus asked 'what does it shut down?' (2011: 8).

My question is if the creation of indigenous studies can entail the relinquishment of all Western elements in indigenous-related research – and it necessarily has to. I argue that Māori rugby and indigenous studies are not too dissimilar cultural phenomena. I like to think of the latter as the indigenization of the Western concept of research. While this phenomenon witnesses creativity, resilience of local epistemes and values, and enfranchisement within knowledge, the Western origins of research and the fact that indigenous studies usually operate within a Western context cannot be transcended. In research, like in daily life, there are no essential definitions of what is Western and what is indigenous, for they have long overlapped.

Some of the elements I have sometimes noticed in Māori contexts – judging internal and external gaze in binary terms (good/bad), perceiving non-indigenous individuals in essentialist terms, fearing to be labelled as 'political' – actually reproduce Western criteria of definition. The picture is just reversed. The issue is that they are some of the discourses that have been used to disenfranchise Māori and delegitimize their episteme. They also overshadow indigenous views which would on the contrary facilitate inclusiveness and collaboration. The emphasis on blatant 'Western' signs eclipses 'the internalization of the symbolic violence' (Andersen 2010: 23) whereby the Europeans have controlled the indigenous, and so conceal the fact that the power of colonialism consists in living in them.

The Māori scholar Borell pointed out that Māori researchers 'must ensure that decolonising projects at a strategic level, do not become *re-colonising* projects at an operational level' (2005: 40). The Māori scientist Hokowhitu advocated that indigenous people and, consequently, indigenous studies should move beyond the concept of decolonization itself, for 'the assertion of Indigenous self-determination in constant referral to the colonising other merely serves to re-establish the neo-imperial colonial power structure themselves' (2010b: 210). In this sense, I think that indigenous self-confidence and self-esteem play an important role. My impression was that individuals who had a more secure identity than others tended to be more conciliatory or open towards anthropology and non-Māori researchers. Encounters and dialogues can contribute to grow such confidence as well as to foster reconciliation with what it is viewed as Western. The multiple conversations/ confrontations Māori and I had – as intellectuals and cultural subjects – certainly breached many barriers creating possibilities for reciprocal understanding and self-questioning as well as sharing.

As a discipline that acknowledges cultural diversity and aims to understand cultures, I believe that anthropology is potentially the Western-defined discipline that can more easily access, make intelligible and legitimate before the Western scientific world the diversity and exigencies of indigenous research. I also view its engagement with indigenous realities as a responsibility. As for Māori Studies, it opens up various possibilities for collaboration, because 'while Māori Studies positions Māori culture, knowledge and values at the centre of investigation and representation [...] it can also be a bridging point between theoretical and disciplinary multiplicity' (Gonzalez 2010: 13). Under the umbrella of Kaupapa Māori research, anthropology and indigenous studies can potentially co-exist as disciplines contributing in their own way to the understanding of Māori issues and realities.

References

Andersen, C. 2010. Mixed ancestry or Métis? in *Indigenous Identity and Resistance: Researching the Diversity of Knowledge*, edited by B. Hokowhitu. Dunedin, NZ: Otago University Press, 23–35.

Asad, T. (ed.) 1973. *Anthropology and the Colonial Encounter*. London: Ithaca Press.

Bensa, A. and Bourdieu, P. 1985. Quand les Canaques prennent la parole. *Actes de la recherche en sciences sociales*, 56(1), 69–85.

Best, E. 1921. The Māori genius for personification; with illustrations of Māori mentality. *Transactions and Proceedings of the Royal Society of New Zealand*, 53: 1–13.

Bishop, R. 1998. Freeing ourselves from neo-colonial nomination in research: A Māori approach to create knowledge. *Qualitative Studies in Education*, 11(2), 199–219.

Borell, B. 2005. *Living in the City Ain't So Bad: Cultural Diversity of South Auckland Rangatahi*. Unpublished MA thesis. Massey University, New Zealand.

Bourdieu, P. 2003. Participant objectivation. *Journal of the Royal Anthropological Institute*, 9, 281–94.

Calabrò, D.G. 2011. *The Indigenization of Rugby in New Zealand and its Role in the Process of Māori Identity Definition.* Unpublished doctoral thesis. University of Messina, Italy.

Clifford, J. 2004. Looking several ways. *Current Anthropology*, 45(1), 5–30.

De Martino, E. 1941. *Naturalismo e storicismo dell'etnologia.* Bari: Laterza.

—— 1949. Intorno a una storia del mondo popolare subalterno. *Società*, 5, 411–35.

—— 1951. Il folklore progressivo. *L'Unità*, 26 June 1951.

—— 1999 [1995]. Notes de voyage, translated by Giordana Charuty, Daniel Fabre and Carlo Severi. *Gradhiva*, 26.

—— 2005 [1961]. *The Land of Remorse: A Study of Southern Italian Tarantism*, translated by Dorothy Louise Zinn. London: Free Association Books.

Evans-Pritchard, E.E. 2004 [1951]. *Social Anthropology.* London: Routledge.

Gonzalez, C.M. 2010. *'Be(com)ing' Ngāti Kahungunu in the Diaspora: Iwi Identity and Social Organisation in Wellington.* Unpublished MA thesis. Victoria University of Wellington.

Gupta, A. and Ferguson, J. 1997. Discipline and practice: "The field" as site, method, and location in anthropology, in *Anthropological Locations. Boundaries and Grounds of a Field Science*, edited by A. Gupta and J. Ferguson. Berkeley and Los Angeles: University of California Press, 1–46.

Hanson, A. 1989. The making of the Māori. The invention of culture and its logic. *American Anthropologist*, New Series 91(4), 890–902.

Hendry, J. 2007. Building Bridges, common ground, and the role of the anthropologist. *Journal of the Royal Anthropological Institute,* 13(3), 585–601.

Hendry, J. and Fitznor, L. (eds) 2012. *Anthropologists, Indigenous Scholars and the Research Endeavour. Seeking Bridges Towards Mutual Respect.* New York: Routledge.

Herzfeld, M. 1992. *The Social Production of Indifference. Exploring the Symbolic Roots of Western Bureaucracy.* Chicago/London: Chicago University Press.

Hokowhitu, B. (ed.) 2010a. *Indigenous Identity and Resistance: Researching the Diversity of Knowledge.* Dunedin, NZ: Otago University Press.

—— 2010b. A genealogy of indigenous resistance, in *Indigenous Identity and Resistance: Researching the Diversity of Knowledge*, edited by B. Hokowhitu. Dunedin, NZ: Otago University Press, 207–25.

Ito, Y. 2004. On Māori Studies as a discipline and its differentiation: From the viewpoint of social system theory. *The Journal of New Zealand Studies in Japan*, 11: 11–23.

James, W. 1973. The anthropologist as reluctant imperialist, in *Anthropology and the Colonial Encounter*, edited by T. Asad. London: Ithaca Press, 41–69.

Kahotea, D.T. 2006. The 'native informant' as Kaupapa Māori research. *MAI Review* 1, Article 1, 1–9.

Kuokkanen, R. 2007. *Reshaping the University: Responsibility, Indigenous Epistemes, and the Logic of the Gift.* Vancouver: UBC Press.

Lewis, D. 1973. Anthropology and colonialism. *Current Anthropology*, 14(5), 581–602.

Maaka, R. 1998. A relationship, not a problem, in *Living Relationship: The Treaty of Waitangi in the New Millenium*, edited by K.S. Coates and P.G. McHugh. Wellington: Victoria University Press.

Mahuika, R. 2008. Kaupapa Māori theory is critical and anti-colonial. *MAI Review* 3, Article 4, 1–16.

Mead, S.M. 1997. *Landmarks, Bridges and Visions: Aspects of Māori Culture.* Wellington: Victoria University Press.

Metge, J. 1976 (revised edition). *The Māoris of New Zealand.* London: Routledge and Kegan.

—— 1990. Te Rito o te harakeke: Conceptions of the whānau. *Journal of the Polynesian Society*, 99(1), 55–92.

—— 1995. *New Growth from Old: The Whānau in the Modern World.* Wellington: Victoria University Press.

Muru-Lanning, M. 2012. *Researching your Own: Māori Scholarship Post-Treaty Settlement.* Paper presented at Ngā Pae o te Māramatanga – International Indigenous Research Development Conference, 27–30 June 2012.

Nadel, S.F. 1951. *The Foundations of Social Anthropology.* London: Cohen and West.

Narayan, K. 1993. How native is a native anthropologist? *American Anthropologist*, 95(2), 671–86.

Nepe, T.M. 1991. *Te Toi Huarewa Tipuna: Kaupapa Māori, an Educational Intervention System.* Unpublished MA thesis. University of Auckland.

Obeyesekere, G. 1997. *The Apotheosis of Captain Cook. European Mythmaking in the Pacific.* Princeton: Princeton University Press.

Ottenberg, S. 1990. Thirty years of fieldnotes: Changing relationships to the text, in *Fieldnotes. The Making of Anthropology*, edited by R. Sanjek. New York: Cornell University Press, 139–60.

Polack, J.S. 1840. *Manners and Customs of the New Zealanders*, Vol. 2. London: Madden and Hatchard.

Reilly, M.P.J. 1996. Entangled in Māori history: A report on experience. *The Contemporary Pacific: A Journal of Island Affairs*, 8, 387–408.

Sakamoto, H. 2011. Researching kapa haka and its educational meanings in today's Aotearoa/New Zealand: Weaving methodologies, perspectives and decency. *The International Journal of the Arts in Society*, 6(3), 57–66.

Salmond, A. 1985. Māori epistemologies, in *Reason and Morality*, edited by J. Overing. London: Tavistock, 240–64.

Saunders, G.R. 1993. 'Critical ethnocentrism' and the ethnology of Ernesto De Martino. *American Anthropologist*, New Series 95(4), 875–93.

Sillitoe, P. 2006. The search for relevance: A brief history of applied anthropology. *History and Anthropology*, 17(1), 1–19.

Smith, G. 2003. *Kaupapa Māori Theory: Theorizing Indigenous Transformation of Education and Schooling.* Paper presented at Kaupapa Māori Symposium. NZARE/AARE Joint Conference, Auckland, NZ, December.

Smith, L.T. 1999. *Decolonizing Methodologies: Research and Indigenous Peoples*. London: Zed Books and Otago University Press.

Sorrenson, M.P.K. (ed.) 1986. *The Correspondence between Sir Apirana Ngata and Sir Peter Buck, 1925–50*, vols 1–3. Auckland: University Printing Services.

Spivak, G. 1990. *The Postcolonial Critic*. NY: Routledge.

Te Punga Sommerville, A. 2011. *Neither Qualitative nor Quantitative: Kaupapa Māori Methodology, and the Humanities*, paper presented at the Kei Tua o Te Pae conference, Wellington, 5–6 May.

Thomson, A. 1859. *The Story of New Zealand: Past and Present, Savage and Civilized*, vol. I. London: John Murray.

Trask, H.K. 1991. Natives and anthropologists: The colonial struggle. *Contemporary Pacific*, 3, 159–67.

Walker, S. 1996. *Kia tau te rangimarie. Kaupapa. Māori Theory as a Resistance Against the Construction of Māori as the Other*. Unpublished MA thesis. University of Auckland.

Webster, S. 1998. *Patrons of Māori Culture. Power, Theory and Ideology in the Māori Renaissance*. Dunedin: University of Otago Press.

Yacin, T., Mammeri, M. and Bourdieu, P. 2003. Du bon usage de l'ethnologie. *Actes de la recherche en sciences sociales*, 150(1), 9–18.

Chapter 4
Hiding in Plain Sight:
Assimilation and the End of Story

Robyn Sandri

When I stand on my traditional Gungarri country the spirits speak to me. I know they whisper with the rustle of the leaves, in whirlwinds and wash of breeze against the long grasses. I sense their presence underfoot in the dust, in the bushes and the creek ripples. I know the spirits call out to me, but I cannot understand their ancient words for my ear, like my tongue knows only English. The old spirits know my blood line and greet me. They sense my belonging. My grandmother's' ancestors lived here since the beginning and until the European invasion. Their shadows speak to me in dreams. They send messages to me with the black and white birds. They are long forgotten knowings, which are familiar. I can make no sense of what they wish me to know.

The First People of Australia survived European colonization and are still here. By 2050 it is estimated half of the northern Australian population will be Aboriginal Australians (ABS 2012). At least 75 per cent of Aboriginal Australians now live in urban and regional areas. Since colonization many do not know their original country and families. Their inter-generational linkages to language, cultural stories, lore, knowledge and traditional connections to the land and kin are lost to the vast majority. It is the existence of culture that affirms a peoples' sense of belonging, identity and self-worth.

This chapter discusses a personal perspective of Robyn Sandri's experiences growing up as a fair-skinned Aboriginal child in Australia and contextualizes it within her doctoral research journey. Robyn grew up in an Aboriginal family during the 1960s with an Aboriginal mother and a Welsh–Australian father. The 1960s were lived under the shadow of Federal Australian White Australia immigration policies as well as the Aboriginal Acts which aimed at mainstream assimilation and integration of native peoples. The fictitious notion of 'Terra Nullius' from first settlement demonstrated that Australia was envisaged by British colonizers from the outset as a white European country. Aboriginal people who could pass as olive skinned, white Mediterranean or Indian heritage did so in order to avoid the harsh prejudices and hold some human rights not accessible to the Aboriginal population. The traumatic impact of colonization processes continue to shape the experiences of daily life for Aboriginals families in this settler society (Sandri 2013).

Indigenous or Indigenist research is an engaged research in that it seeks to hear the authentic voice of the native. It claims shared space within the traditions of academia. Indigenous experience and voice is not silenced or 'Othered' by the Western academic institutions. To 'Other' identifies those considered different from the mainstream culture, it is used to reproduce positions of domination and subordination. Linda Smith (1999) argues that Indigenous research developed at the intersection of post-colonial theory (including theorists such as Fanon, Spivak and Said) and decolonization theory. Post-colonialism was traditionally considered a time period of history of 'handing back' colonized states by supreme powers to original inhabitants. Decolonization theory, unlike post-colonization theory, acknowledges that the colonists have not left. It calls for 'socially just' recognition and respect for First Peoples. This is contentious ground for First Nation academics as research and doctorates are by nature Western. As dominant Western thought and ways of knowing pervade within the academy, securing an equal space or voice is difficult. It implies changes within the systemic academic authority.

Robyn recently completed research which relied on emerging Indigenist research methodologies that were informed by international native researchers including Archibald (2008), Rigney (1999) and Smith (1999). Indigenous research includes the use of auto-ethnography so that the voice of the Indigenous researcher is included. Narrative research is used so that the participant's voice and power are more equally balanced with the researcher. Indigenous ethnography replaces the use of classic ethnography because it originated as a way for the Western researcher to explore and examine the 'exotic' Other (Patton 2002). Indigenous ethnography explores the native voice within its own perspective, context and experience. The Indigenous instruments allow the story to be unfiltered by the Western interpretative lens. The storied data is not 'Othered'. Indigenist research offers experiences and voices of colonized people to be added to the historical accounts of European colonization. It offers a more balanced and truthful telling. In my family our Aboriginality was kept secret from the outside world. I did not fully discover or understand that we were Aboriginal until I was in my twenties and lived overseas. In 2006 I returned to Australia to undertake Indigenous doctoral studies. I was unsure of my native identity but claimed Aboriginality as it was my heritage and the focus of my study. I was neither an 'insider' nor 'outsider'. I questioned my sense of 'self' and 'identity'. If self 'refers to the subject and the social values acquired by an individual to position self within a construct' (Kumar 2000), I was not Aboriginal. Nor was I fully European Australian as many of the norms in my upbringing were Aboriginal but not labelled as such. When I contemplated my lived experience it did not feel 'hybridized' as half of my heritage was not celebrated. It was concealed with shame. I realized that what I shared with Australian and other international Indigenous peoples was a colonized position. As an academic I identified as a 'colonized Indigenous person' as that best described my worldview and life experiences within Australia.

What I was unprepared for was the assigned and entrenched stigma of Aboriginality within the Western academic community. In my doctoral thesis, I considered the historical and contemporary difficulties for Indigenous families engaging in academic and schooling systems. All families in the study reported incidents of racism, inequality and exclusion. During my research, I discovered my personal and racialized history and the inter-generational trauma experienced by people who live without their sheltering cultural stories, lores and beliefs.

The history of the settlement process was largely unrecorded from the Aboriginal perspective. It has not been considered by the mainstream in terms of trauma, post-traumatic stress or inter-generational and trans-generational notions of trauma transference. I found this to be evident in current interactions with authority figures such as teachers and other officials in government institutions. The impact of inter-generational trauma on Aboriginal children in schools is not considered in theory of practice within schools. Poor Aboriginal schooling outcomes continue to be determined as a family or individual problem (Malin 2003). This is an example of blaming the victim which alleviates any systemic responsibility.

The women of my maternal Kooma-Gungarri line lived in their country in the Maranoa district of South West Queensland since the beginning. The Maranoa 'mobs' or tribes are desert people. They are known as 'brown-water' mobs. Traditionally, tribal groups were identified across Australia by their water source as it differentiated geography and country as well as ways, customs and character. Freshwater people were the tribes living on the coastal rivers. The Undumbi or saltwater people lived along the coastline and off the mainland on the islands. Salt-water people lived with a year round abundance of seafood, and land animals such as emu and kangaroo to hunt. Brown-water people were supposed to have harder living conditions, although my grandmother's tribal group lived in lands where kangaroo and emu are plentiful. They were also a fishing people with wide inland rivers to supply fish. They had a variety of native bush foods for fruit, vegetables and medicines. While I've never lived on my country, I have this innate sense of knowing about it. I do not fear the lizards, snakes and bird life but I am not comfortable with sea creatures. I do eat anything from the saltwater. I like the sense of belonging on those long dry plains that run on to the horizon and the emu and kangaroo that fill the roads at sunrise and sunset. My dreamings and stories are from those places although I only hold a sense of them.

I do not know the old cultural stories and everyday ways of tribal people, so I do not know what it means to a traditional Australian. The language, beliefs and lore were already colonized out of my family when I was born. Language, knowledge, intelligence and memory are all intertwined, and the old lore is lost forever when the language is gone. My old people tell me that I am related to the emu, the possum, the brown snake and the wind. It deeply saddens me to not know what that means.

There really is no way to appreciate the depth of that interconnectedness to your country and every other thing unless you are Aboriginal; and maybe you even need the old language in your head to fully know it. The Aboriginal creation spirits, known as the Wagalak sisters travelled the land following the rainbow snake. The snake made the landscape then they carried the people, their lore, language and knowledge in their sacred dilly bags and placed them in the country where they belonged. The people connected to their country in every way. They were related to each other, to the land, the birds, animals, trees and plants. They were related to the weather, the air, and the night sky. There are no words in the English language to describe this connection. The deeper and sacred levels of culture are missing now. The spirits of the country still hold their stories but few have the language to know them now.

The stories were known by the tribe since the beginning, and those stories wrapped around the people like a protective mantle. They carried and affirmed the culture, and taught the traditional sacred ways as well as the common everyday ways from one generation to the next. Parents taught the children, and families held knowledge, values and beliefs even if they were not spoken aloud. They were innate and implied. If children are removed from their families they are removed from their cultural protections and from their cultural teachers (Halloran 2004). The contribution of early anthropologists such as Norman Tindale, Charles Mountford, Frederic Wood Jones, Thomas Campbell and Robert Pulleine were to record and hold some knowledge in their pioneering ethnographies. Ironically, following colonization and disruptions, some of the anthropologists' records hold the only cultural knowledge at times. The Tindale tribal map of Australian tribal groups, while controversial and inaccurate, is used in native title claims today.

In all cultures, cultural maintenance allows people to know who they were, and how they matter in a capricious world. Its existence depends on ongoing cultural teaching. An oral culture is most vulnerable to loss because life is fragile. It gave Aboriginal people their self-worth and self-esteem. It offered them a space where they belonged and had obligations and where they mattered. When 'stolen children' were placed in non-indigenous homes and institutions they knew instinctively they were not European, but they did not know how to be Aboriginal. If the culture is destroyed, people live in a space of cultural trauma (Aitkinson, 2002)

Aboriginal people were forbidden by settlement processes and in missions from telling their stories and speaking their languages. Initial attempts to disperse Aboriginal people were highly successful. Dispersal acts did not openly encourage but condoned the removal of Aboriginal people from settlers' properties. Tribes were forcibly removed from their land as settlers fenced their properties. What European officials called 'dispersals' were referred to by Aboriginal people as the 'fencing times'. They are also known as 'the killing times' as settler led massacres were common.

Tribal groups were removed to missions. The state governments assumed that Aboriginal people would not survive modernization and colonization, so the missions would provide a place for them to die out. While people were treated appallingly on missions, many were protected from violent deaths and many did survive (Harris 2003). Other Aboriginal people were employed by the property owners or town people to work as domestics could be exempted from removal. They typically lived in small tribal or family groups in town camps called 'yumbas'.

My great grandmother Alice and her family lived on the Bollon yumba. A yumba is a permanent town camp or fringe dwelling which was outside the boundaries of the town. They could be built on the river, the cemetery or the town garbage dump. The houses were built from whatever materials were available, such as oil cans hammered flat. Alice died in childbirth in 1933 when she was 30 years old. She and the infant were buried in an unmarked grave in the cemetery next to the yumba.

The local police man, as protector of Aboriginal people, was sent to investigate her death. In his report, which I located in archives recently, he wrote that the family was clean, well fed and the children were sent to school. The family was granted an exemption by the police man. This meant they would not be removed to a mission. Regardless of the exemption, my grandmother Amy and her brother Tommy were taken with other children in the nearby town of St George one day when they went shopping. There was a systematic and general round up of Aboriginal children as it was seen they needed to be removed from their cultural teachers if Australia was to be a European nation.

Governments had such power over Aboriginal families in the 1900s because all Aboriginal children were born as state wards. Parents had no rights over their children. Many babies were removed at birth in hospitals, others taken at any time. Typically, it was the police who removed children from their families just as it was usually police who took on the paramilitary role of removing families from land and in dispersing them.

The government deemed this action necessary and urgent as many Aboriginal children living on the yumba's were light skinned enough to look European, yet growing up culturally as Aboriginal Australians. Recently, Gungarri people in the Maranoa area told me that my grandmother Amy was sent to Taroom mission, but there is no record of her in the mission register. She was 11 when removed; I suspect she may have been sent out to work on a property. She did not speak of the experience ever, however once she spoke of being a cook in a hotel in far Western Queensland when she was 14.

At 21 the Queensland government granted my grandmother a 'Certificate of Exemption'. This meant she was free to live as any European woman, provided she kept to the terms of her exemption. These were that she did not speak her own language, did not interact with any Aboriginal people including her family and did not drink alcohol. As far as I know, she never did any of these things including associate with her own brothers and sisters. My mother tells me Grandma wrote a letter to her sister and arranged for an 'accidental meeting' at a railway station. For 10 minutes two sisters, one olive skinned and the other black stood together on the Roma railway station. I know of no other time they met.

My grandmother married an Australian man with a Greek father and English mother. Regardless of her ethnic heritage, my grandmother reinvented herself and her family as Greek. I do not know what life was like for them, but the need to conceal their Aboriginality was paramount. When she gave birth to a very dark skinned baby boy, the family claimed he was an adopted 'bush baby'. The family lived what is known now as 'passing' or 'false whiteness' (Paradies 2006). In my own family as in many, the fear of discovery was so real not even the children were told of their heritage. Many other Aboriginal Australians have had the experience of discovering as an adult that they are Indigenous Australians. Historian Bruce Pascoe (2008) lived such a life, and when his traditional heritage was revealed to him he went in search of his cultural roots. All that was left of his at least 40,000-year-old heritage was a list of 14 nouns from the tribal language.

My mother would not talk to me about it. Her denial was so ingrained and her false life so normalized that she could not. I came home to Australia and, after speaking to my relatives, put the threads of my family's story together. I drove inland to Gungarri country passing a sign that read 'Welcome to the Outback'. We were desert people and the roads were littered with kangaroo and emu road kill. I found the extended family and once they heard of my heritage they welcomed me. They were obviously Aboriginal people. They had dark skin and Aboriginal features so they had not had any opportunity to 'hide out' from the lack of rights and racism. These were the children and grandchildren of our common great grandparents.

The black family welcomed me home, but nothing about it felt like my home. They looked out at me from their brown eyes to me with my fair skin, but they had other fair family members known as 'fair skinned Aboriginal people'. In a small bakery café in the Queensland town of Mitchell, Queensland I spoke with a man who was my cousin. He would take me up to the Carnarvon Gorge which was central to the Gungarri 'Dreaming stories'. There were ancestors still there; wrapped in bark and their possum skin cloaks and deep in the burial caves. I thought it likely that the DNA from the ancestor remains would likely match mine in some measure.

He told me to walk on country with my shoes off so I could become known to the spirits of the land and ancestors, and they could talk to me. I found family in every conversation with strangers in every café and store. I discovered one of my girlhood school friends was my blood cousin. I was directly related to most in town. I wish I could say this gave me comfort and a sense of belonging but I felt no connection. What had been shadows in my childhood was now reality. Stories of killings and mission injustices became my family heritage and my stories too. Acts of colonization were no longer other families' stories.

The stories brought a sense of knowing to me. It was as if they were reminding me of things I knew once but had forgotten. My intuition and dreams were now understandings which no one ever taught me. I read the world and listened to the wind all of my life. I treasured the birds and wondered what stories they were delivering but I never spoke of these things. I believed these ways were my eccentricities, but I found now they were legitimate ways of knowing in the holistic Indigenous construction of knowledge. I learned that not all knowledge is Western and siloed away from the landscape in books and databases. My black family had welcomed me, but my so called Greek family shunned me. I was disturbing their created identities. They had invented lives of Greek sailors, worked in Greek cafés, cooked Greek food. I was told by an aunt that I was picking over the bones of her mother. I had wanted to find where I belonged, but in truth I found I was even more disconnected. I could not identify with the life of the traditional families, and my own immediate family closed ranks against me. Even my professional home did not want to know about my new understandings of Aboriginal Australian life and the impacts of colonization and ways of knowing. A head of school told me, that if I did not give up my 'Aboriginal interest', I could not 'have a future at this university'.

Working in the Aboriginal Context

My reason for returning to Australia was to undertake doctoral studies. I was working for the state education provider, Education Queensland, establishing Indigenous playgroups north of Brisbane. Aboriginal children were not transitioning into school successfully, and the playgroup initiative was to create some informal pathways to encourage families into schools as well as offer some academic socialization to the children. Both mother and child attended playgroup, and the goal was for parents to see safety and partnerships in schools as an option for them. The playgroups were purposely located in the schools so Aboriginal families could feel culturally safe entering schools. I decided to use the playgroup experiences in my research.

I learned more of the challenges of Aboriginal life and 'not belonging'. I assumed I would offer playgroups and perhaps advertise with a poster or local newspaper advertisement, and the families would attend. No-one came. Aboriginal families typically do not send their children to schools until they are legally required to do so at the compulsory age. I did not realize the sense of exclusion and unwelcome that Aboriginal families experience in schools. I found this belief was trans-generational.

I realized I was part of the problem. One of the Aboriginal grannies told me they thought I was 'too posh' with my nice clothes and how I spoke. After all, I was white and educated. I was working with a group of black welfare families. I located a local community woman as an Indigenous teacher aide to run the playgroup. The families easily related to the new leader and felt more culturally safe in the school. It evolved with time into an Aboriginal community social group, not a government program.

I sat on the ground at playgroup and told the mothers my story. I told them who I was, who my families were, where my country was and my life 'in hiding'. I thought they may identify with me when they saw we did have Aboriginality in common. I had assumed they knew about their Aboriginality, but I was wrong. The families saw that I was open to them, but despite our shared heritage I remained an outsider. I had not faced the racism and the construction of Aboriginality which shaped their lives. I never claimed insider status with this group, as that is something only the group can assign. Over a two-year period they began to trust me and shared their stories with me. This led to my engaged researcher role with the playgroup families.

The Research Quest

I had developed trust with the playgroup and decided to request its use as my research site. Indigenous research should grow from an issue that impacts Aboriginal people, not a researcher's idea for a doctoral thesis. I was imbedded with the playgroup for another three years to undertake this research. As Indigenous researchers such as Martin (2008), Rigney (1999) and Smith (1999) attest, Aboriginal people are tired of being the subjects of research, especially when they see so few positive outcomes. It was important for me as a researcher to do no harm, to be respectful and offer benefit to the community. My observations of how difficult it was for Aboriginal families to come into schools startled me. I was curious as to what caused their disengagement from schools and from creating educational partnerships with teachers and schools.

The playgroup was developing into a space of cultural safety within the school for the Aboriginal families. Engagement between parents and teachers, and community and schools are known to be key determinants of successful schooling and life trajectories for Aboriginal children (Malin 2003). I noticed real emotional difficulties such as anxiety and avoidance from many of the families when they were called upon to interact with teachers, principals and school administrators. I had seen that one playgroup grandmother refused to go into the school at all. If the teacher wanted to speak with her about her older children, they would have to go to her home. Another would not go to school appointments and had to be accompanied by the Aboriginal community liaison officer. These experiences demonstrate how Western schooling is a cultural mismatch for Aboriginal families and children. The also demonstrated how emotional and traumatic schooling memories travelled across generations. Eventually the families came into the playgroup at the school happily and engaged with each other in positive ways. Significantly, I noticed the families presented as passive and disengaged if white researchers or observers attended. My quest was to understand across generations what the experiences of school had shaped their disengagement.

Methodology

I own that at the beginning I was influenced by my Western scholarship and training in Australian and American Universities. I designed my research process with the assistance of my non-Indigenous supervisor as a qualitative ethnographic study using interviewing techniques with predetermined questions. I came to see that Western research was innately othering. Indigenous worldview remains the point of difference. Indigenous research paradigms undertaken by Indigenous researchers are within the common lens of heritage, history, social and cultural lifeworld. By contrast, Western researchers view the Indigenous context though their own lens which by nature is othering.

The university Ethics Committee took 18 months to grant my approval. It was a process designed for Western models of research and did not have the capacity to work outside of that context. When they read I was undertaking research with an Indigenous community they wanted approval from the local community council or health service. I explained that this was not a rural, remote Deed of Trust community (a Deed of Trust community is typically an ex-mission which is now administered by an Indigenous council). I was undertaking urban research in a small town near the major metropolitan area of Brisbane. At this time the vast majority of Aboriginal families live in suburban and urban communities. There was no designated Aboriginal entity, as would be found in remote communities. When I did submit a letter from the traditional owner and elder, the ethics committee wanted a document on official letter head. She was an individual in the suburbs surviving on a government pension, not a registered or funded organization.

At the oral presentation and defence of my original research proposal an Indigenous panel member alerted me to how Western my methodology and literature remained. I read more broadly about Indigenous research, methodologies and knowledge. My supervisor felt I needed strong Western design, theories and literature to validate a doctoral thesis. I was positioned in a challenging space, balancing the university's need for Western literature and methodologies with Aboriginal lived reality. Systemically, the university did not have the capacity to offer me an Indigenous supervisor and I was totally unprepared for the challenges.

I found my initial attempts to work with the families and gather interview data was unsuccessful. Parents, who were friendly with me in an informal context, were uncomfortable with my new official role of university researcher. They preferred not to separate from the support of the group, or from their young children to speak with me in the adjoining office. Families became politely distant, and pleasantly non-responsive to me. The mothers responded to my initial data gathering to say that school had been a pleasant experience and they had difficulties doing the work. No-one reported instances of racism or othering problems at school. This data supported the Western assumptions that Aboriginal pupils were 'the problem'. The mothers responded to my questions in non-contesting ways; not drawing any attention to their negative experiences.

From the outset it was apparent that my research process and design were Western and innately othering. I came to understand this was an Aboriginal strategy of invisibility. They had employed invisibility as a survival technique. In frustration, I decided to simply attend playgroup weekly, so the newer families would get to know me and in time reveal their real experiences. I would add patience to my research tools. Had I not had shared colonizing experiences and time within the playgroup families I may not have recognized this as strategy, and reported data which reinforced deficit views. When Linda Smith (1999) called for Indigenous researchers to first decolonize the mind, I came to understand what she meant. I worked with the advice of Elders and re-configured my research plan to a design which allowed the community issues to emerge. I also had to do so in a way to honour the Western university tradition of a doctoral thesis. I sought to balance research styles. There were emerging new ethnography and narrative methodologies which could be well used in both domains (Clandinin 2007).

The Invisibility of Western Assumptions in Research

As I had designed a study across generations, I had always planned to interview my storytellers in age order, thus locating each generation within its time period and colonization experience. This design revealed how colonization had disrupted families. I planned to interview across three or four generations within each family. I could not locate that many generations in any one playgroup family. Many of the playgroup mothers did not know their parents, grandparents or traditional tribal country. It was difficult to find older Aboriginal people as so few Aboriginal people live into old age. According to the Australia Bureau of Statistics (2010) Aboriginal people have a life expectancy of at least 17 years less than other Australians. In personal communication Aboriginal medical doctors report more startling figures: in Queensland the mean age of death of Aboriginal men is 57–58 years.

As I began to hear their stories, I discovered that all of the families were impacted by the assimilation policies which removed children from their families. I redesign my study as trans-generational rather than within family groups or inter-generational. Nothing about my original Western design existed as it did not reflect the world of Aboriginal Australians; it did not accommodate the reality of colonized Indigenous lives. I saw how important it was for them to remain invisible.

Initially, the playgroup mothers preferred to be interviewed within the group, however over time they would seek me out to give me more detail or tell me other stories. The storytellers did not want to be identified in the research. I honoured this request and did not name them or the town. I realized the storytellers did not like the use of audio visual devices. I had planned to record their stories and have them processed by a transcription service. Instead, I listened to the stories as a conversational partner, and just made a few notes at the time. This technique allowed the story tellers to be open. Immediately afterwards I wrote out the stories fully. I then word processed the notes to the story and revisited the storytellers for their review.

As I had removed the predetermined questions, I asked the storytellers to talk to me about their schooling experiences. I explained I wanted to understand if their school history affected their attitudes to school and education. The Elder and grandparent storytellers were less concerned about being identified. In fact, two of the Elders had penned autobiographies and they were accomplished storytellers. They told their stories with a long timeframe, knowing how their own experiences had shaped their long lives. While the young mothers were reluctant to tell me their stories, the Elders were so forthcoming that I had to turn away offers. I was approached by people from all over the state wanting to tell me their life stories. The Elders often spoke to me of their parents and families so I found I was collecting 100 years of Indigenous family history. This was a small study of only five storytellers for each generation so it does not represent all of Aboriginal Australian history.

Two of the Elders were 'mission bred', which is the term in common usage by Aboriginal people to designate those who grew up on missions. One family was forcibly removed to a mission. Another Elder in the same mission had moved with her family to escape the fear of being killed or starved off their traditional country. Both missionized storytellers spoke of awful and inhumane schooling. One of them, Uncle Henry, spoke of how the little girls were whipped on their bare legs. Another, Aunty Ruth, told me how a cat of nine tails was used on them regularly, 'We were treated like prisoners, but we were only little children', she said. She had published the story of her childhood mission experiences (Hegarty 1999). The most poignant image of mission schools was from Uncle Henry. He described how at the end of the school day, the children were forced to lick their slates clean. The image I see is of little Aboriginal children forced to swallow white words.

Another two of the Elders spoke about having to pass as white to have life opportunities. One family reconstructed itself as an Italian family and used an Italian name. Another Elder spoke about how she planned to have the church fund her education and send her to the city to train as a teacher. Even when qualified, she was not allowed by authorities to work in mainstream schools and was sent to the Torres Straits to teach black children. The final Elder storyteller spoke about life in a mainstream school. She said that 'the teachers equated intelligence with skin colour'. She was top of the class academically, but she was not recognized for her scholarship or called upon to answer questions. She experienced racism and was often excluded from lessons. She decided that if she remained silent and invisible in the classroom, she was not excluded from lessons.

While the Elder generation spoke of horrific experiences, I was unprepared for the stories the next generation of storytellers were to tell me. These were the stories of the stolen children who were in their fifties now. They called themselves 'us taken away kids'. While the Human Rights and Equal Opportunity Commission had investigated the practice and published their findings in 'The Bringing them Home Report' (1997), I found it sanitized the reality of their lived experiences. The real horror was what happened to the children after removal. Of the five I interviewed, only one was placed in a home which offered him respect if not love. He was removed at birth from a Sydney hospital. The mother was released without her son. She came every day to beg for her baby. She was never allowed in to see him, and committed suicide by throwing herself under a train. All of the other interviewees spoke of sexual abuse. This occurred whether they were placed in institutions or private European Australian homes.

However, while family lines were destroyed, and Aboriginal culture not transmitted, the children were neither Aboriginal nor white. They had no secure cultural identity. These policies saw Aboriginal culture destroyed. Consequently, few contemporary urban Aboriginal people have little knowledge of their tribes, language and cultural lore. I also realized no Queensland educational policy considers the impact of the trauma experienced by these families.

Over the time I was immersed in the playgroup I became one of 'the mob'. I was assigned the role of advocate and storyteller, and I retold the stories I had gathered. The youngest current day mothers who initially told me pleasant stories began to share their real experiences with me. The most enduring themes were of exclusion and racism. The mothers insisted I hear the story of one non-indigenous mother married to a traditional Aboriginal man. She had four indigenous children. According to the Australian Bureau of Statistics (2012) the contemporary and upward trend is for Aboriginal people to marry non-indigenous partners. Considering this trend her story was relevant to this study. When she sent the children to the religious school she attended happily as a child, issues relating to Aboriginality emerged almost immediately. She felt any issue that occurred was constructed as a problem of their Aboriginality.

The Silence of the Children

I had planned to include the children as a fourth generation of storytellers. The playgroup hub became a space of cultural belonging in the school, and just as it was usual for Elders of the community to drop by, the older children came by too. I did not identify a 'child' sample; I just said that any of the children who would like to offer me a story about school or playgroup were welcome to do so. When I attempted to interview the children, I received nothing but polite smiles. I used classic art and drawing techniques as well as puppets to talk about school. The children enjoyed the games, especially the puppets, but revealed nothing noteworthy to me. They were silent, just as my own childhood silence was imperative. It was a technique of invisibility.

I pondered why the youngest generation was reluctant to speak to me when I was known to them. I am not sure that European Australian children would have been so reluctant. If I were a non-indigenous researcher on a time schedule I may have constructed the silence in deficit ways. I may have surmised the children lacked social and conversational skills, or that they had language delays. I felt deeply but instinctively that there was something going on here that really mattered. I returned to the Elders stories of schooling experiences. They all said they felt unwanted at school and they learned to strategize invisibility and silence to gain an education.

Child welfare and child protection services remain an enduring issue for Aboriginal families. Aboriginal children remain over- represented in current removals of children into state care. They are 10 times more likely to be removed than other children (Australian Institute of Health and Welfare [AIHW] 2013). Most are removed on the designation of neglect, which is directly related to poverty and Aboriginal Australians live marginalized and impoverished lives on any measure (AIHW 2013). Through the playgroup years, I continued to witness how reluctant the families were if white researchers, visitors or child safety officers were present. The families would not come if they knew authorities or strangers were attending, or disengage by sitting in silence at the back of the room and leave early. There were big risks associated with authority figures like police, teachers and social workers.

The playgroup children were not interested in me as I was simply another adult at the sessions. It did not give me an insider status to them, even if the mothers offered it. The parents may have understood that, but I suspect to the children I appeared just like any teacher in the classroom or the principal in the administration office. I looked like the enemy, or like the sort of person who might take them away. Even when the Aboriginal mothers encouraged the children to speak to me, they did not. Their eyes would dart from me to their mothers looking to be released to run outside to play with the bikes or cars or climb on the play equipment.

In the study, each generation affirmed experiences of discrimination, racism, and low expectations at school. They all lived with fear and anxiety. According to psychologist Michael Halloran (2004) people need cultural security to cope with life. When denied, Aboriginal people have endured cultural genocide. They survive in a space of cultural trauma which manifests in behaviours and illness similar to post traumatic stress (Halloran 2004). Since colonization, Aboriginal culture remains in remnants to a lost people.

Policies since the early colonial designation of Terra Nullius have underpinned discrimination. The Australian Constitution continues to allow discrimination against Aboriginal people. It affects all Indigenous Australians (Aitkenson 2004). Professor Marcia Langton (personal communication 2011) calls for the words 'race and ethnicity' not to be applied to Aboriginal people. It is a time for healing and human rights for all Australian people. Nowhere is this more apparent in the health, welfare and education of Aboriginal children who are denied chances equal to that of other Australian children. The playgroup created change for the community in ways I never expected when I began it. The hope is the playgroup families came together in the process of reconstitution or building an urban tribe, gaining power in the process (Foucault 1990). The mothers are creating the new Aboriginal tribe with their stories of change for their families and children. In Australia increasing numbers of people are identifying alongside the growing number of white Aboriginal people. The term 'fair skinned Aboriginal people' was coined by the Aboriginal community as many families have fair members. The real change is coming from the ground up, not from the government down. The First People are building tribe, exercising power and writing new stories about resilience and resistance and claiming power to create change.

Although my research had begun to consider the transition of Indigenous children into mainstream school I found the research touched on my story. It was an emotionally demanding task. I heard of murder, sexual abuse, hounding and mission life I knew to be associated with my own family. I realized what colonization had meant for my mother's family and understood my forbears' lives. I understood how important it was to be invisible, or pass for white in Australia.

References

Archibald, J. 2008. *Indigenous Storywork: Educating the Heart, Mind, Body and Spirit*. Toronto, Canada: UBC Press.

Atkinson, J. 2002. *Trauma Trails Recreating Song Lines: The Trans-Generational Effects of Trauma in Indigenous Australians*. Melbourne, Australia: Spinifex Press.

Australian Bureau of Statistics and Australian Institute of Health and Welfare 2008. *The Health and Welfare of Australia's Aboriginal and Torres Strait Islander Peoples* (ABS Catalogue No. IHW 42). Canberra, ACT: Australian Government Publishing Service.

Australian Bureau of Statistics (ABS) 2012. *National Aboriginal and Torres Strait Islander Social Survey* (ABS Catalogue No. 4714.0). Canberra, ACT: Australian Government Publishing Service.

Australian Institute of Health and Welfare (AIHW) 2013. *Australia's Health 2013: Australia's Health Series*. Canberra, ACT: Author.

Clandinin, D.J. 2007. *Handbook of Narrative Inquiry: Mapping a Methodology*. Thousand Oaks, CA: Sage.

Closing the Gap 2009. *Closing the Gap Clearinghouse Annual Report 2009–2010*. Council of Australian Governments (COAG) Australian Institute of Health and Welfare and Australian Institute of Family Studies. Canberra: Australian Government.

Foucault, M. 1980 [1966]. *The Order of Things: An Archaeology of the Human Sciences*. (A.M.S. Smith, Trans.). London: Tavistock.

Halloran, M. 2004. Cultural maintenance and trauma in Indigenous Australia. *Murdoch University Electronic Journal of Law*, 11(4). Retrieved from: http://www.austlii.edu.au/au/journals/MurUEJL/2004/36.html, accessed: 12 April 2012

Harris, J. 2003. Hiding the bodies: The myth of the human colonisation of Aboriginal Australia. *Aboriginal History*, 27, 79–104.

Hegarty, R. 1999. *Is That You Ruthie?* Brisbane: University of Queensland Press.

Human Rights and Equal Opportunity Commission (HREOC) 1997. *Bringing Them Home: Report on the National Inquiry into the Separation of Aboriginal and Torres Strait Islander Children from their Families*. Canberra, Australia: Human Rights and Equal Opportunity Commission.

Kumar, M. 2000. Postcolonial theory and cross-culturalism: Collaborative 'signposts' of discursive practices. *Journal of Educational Enquiry*, 1(2).

Malin, M.A. 2003. *Is Schooling Good for Indigenous Children's Health?* (Occassional Paper Series: 8). Darwin, NT: The Cooperative Research Centre for Aboriginal Health.

Martin, K.L. 2008. *Please Knock Before You Enter: Aboriginal Regulation of Outsiders and the Implications for Researchers*. Brisbane, Qld: Post Pressed.

Paradies, Y.C. 2006. Beyond black and white: Essentialism, hybridity and indigeneity. *Journal of Sociology*, 42(4), 355–67.

Pascoe, B. 2008. Paper trail. *Voice of the Land*, 36, 14–16. Retrieved from: http://www.fatsilc.org.au/voice-of-the-land-magazine/vol-30–39–2005–2009/vol-36-march-2008/14-paper-trail-by-bruce-pascoe, accessed: 15 April 2012.

Patton, M. 2002. *Qualitative Research and Evaluation Methods* (3rd ed.). Thousand Oaks, CA: Sage.

Rigney, L. 1999. Internationalization of an indigenous anticolonial cultural critique of research methodologies: A guide to indigenous research methodology and its principles. *Wičazo Ša Review*, 14(2), 109–21.

Sandri, R. 2013. Weaving the past into the present: Indigenous stories of education across generations. Unpublished doctoral thesis. Queensland University of Technology, Brisbane.

Smith, L.T. 1999. *Decolonizing Methodologies: Research and Indigenous Peoples*. London: Zed Books.

PART II
Problems of
Representation and Rights

Chapter 5

The Promises and Conundrums of Decolonized Collaboration

Emma Cervone

The title of this edited volume hints appropriately at the temporal dimension of the discussion on engaged anthropology and indigenous studies. This dimension is the point of departure of my discussion on the implications of practicing a form of anthropology that has opted for politically engaged methodologies in the production of anthropological knowledge about indigenous societies. My goals are threefold: firstly, to underline the situated nature of such debate by revisiting the most salient moments that have led to the affirmation of ethical and political engagement in anthropological practice (Haraway 1988). Secondly, to suggest ways in which engaged anthropology and indigenous studies can contribute to larger debates about epistemology and methodology as they develop in the discipline. Lastly, to consider the issue of decolonization and suggest two major implications for the decolonization of an anthropological practice premised on collaboration with indigenous actors, scholars and communities. My reflections are based on my eight years fieldwork experience in Ecuador during which I worked with different indigenous and non-profits organizations, and on my present experience both as a member of the anthropological academic community in the USA and as a researcher who continues to engage with collaborative research projects with indigenous organizations. While my experience in Ecuador made me keenly aware of the complexity of indigenous studies in contemporary transnational and national contexts and the challenges these pose to the conducting of anthropological research, my academic experience in the US urged me to revisit older debates on objectivity, subjectivity and positionality in the social sciences to affirm the validity of what for me in Ecuador had already become a legitimate methodology and practice, that is an anthropological practice that we broadly identified as 'engaged' or *comprometida*.[1]

---

1 I analyze this debate extensively in the (2007) article 'Building Engagement: Ethnography and Indigenous Communities Today,' in *Transforming Anthropology*, vol 15 (2), 97–110, 2007, John Wiley & Sons, Inc. This chapter contains some excerpts from this article.

Historical Overview

The temporal dimension of the debate poses the questions of why and under what circumstance discussions on engagement became relevant in anthropology. As early as the 1940s Margaret Mead underscored the professional responsibility of the discipline in matters of public interest and advocated a visible role of anthropologists (Mead 1943). Yet, the critique to anthropology's colonialist roots started at the end of the 1960s when many anthropologists from both developed and developing countries advocated for a more politically aware anthropology that questioned power relations and took a stand in favor of the oppressed. These were the years immediately following anti-colonial struggles, in which leftwing parties and ideologies in different parts of the world were questioning power relations and economic exploitation both at national and transnational level. At this time Marxist-inspired intellectuals such as Günter Frank elaborated the dependency theory which saw in development and aid a new form of colonialism informing postcolonial geopolitics. Researchers in developing countries back then, including anthropologists, felt compelled to advocate the decolonization of anthropology (see Stavenhagen 1971). In the United States, *Current Anthropology* published a special issue in 1968 in which Berreman, Gjessing and Gough argued for a more politically grounded and engaged anthropology, one that would challenge the myth of a value-free social science, that would destroy the 'sacred Ivory Tower of a science for a science's sake' (Berreman, Gjessing and Gough 1968: 394). Important contributions to the debate on power differentials in anthropological practice and its consequences for the production of knowledge also came from feminist scholarship, which criticized gender bias in the social sciences and its epistemological flaws. By emphasizing the role of 'Man the Hunter,' for example, male-biased analyses of forager societies downplayed the important role that women had in the productive sphere (Slocum 1975: 49).

The push towards such critical reflection came also from disenfranchised groups and from indigenous activists. In Latin America for example, the participation of indigenous people became pivotal in advocating a more egalitarian and horizontal relationship between researcher–researched. In the Declaration of Barbados of January 1971, indigenous actors, progressive members of the Catholic Church, and some social scientists denounced the participation of anthropologists in structures of oppression and domination (IWGIA 1971). The emergence of engagement in anthropology responded to the demand of indigenous activists and other social actors who were urging for a clear positioning of every researcher. All such critiques came to question the construction of 'otherness' as anthropological knowledge and methodologies had defined it.

This debate is to be understood as politically and ethically situated within a specific historical and cultural context. With the collapse of the Soviet Union epitomized by the destruction of the Berlin Wall and the consequent demise of communism and socialist ideologies, the discipline turned to what has been defined as the post-modern critique to reason and empirical truth. In the US, seminal works such as *Writing Culture* critiqued anthropological texts by deconstructing modernist anthropological discourse of truth. Instead, the authors of *Writing Culture* advocated a more experimental, reflexive and subjective methodology debunking the illusion of objectivity in anthropological knowledge. In the field of indigenous studies, this climate of disciplinary critique led to anthropological inquiries that aimed at giving a voice to those who had been represented by decades of anthropological ventriloquism. Curricular reforms in US academies became paramount to incorporate testimonial narratives based on indigenous voices as important tools of inquiry (Huizer and Mannheim 1979, Morin and Saladin'Anglure 1997, Muratorio 1991, Pratt 2001). [2]

The case of Rigoberta Menchú, the impact of her testimonial narrative in the peace process of Guatemalan civil war, and the controversy that generated in the US, stand as an illustrative example of this process.[3] The visibility of these political actors within academic settings, together with the advocacy role assumed by many anthropologists in defense of indigenous groups afflicted by political violence and violation of human rights, was pivotal in creating a space for discussing the decolonization of anthropology and indigenous studies. The collaboration between indigenous actors and anthropologists also helped in fostering the process of both formation and visibility of many indigenous intellectuals in different academic and non-academic settings.

The situated nature of the discussion on engagement and indigenous studies opens the door for highly controversial debates in anthropology. The most exemplary recent controversy around this form of anthropological practice took place in US academia at the end of the century when the journalist Patrick Tierney (2000) published an exposé of American anthropologist Napoleon Chagnon who worked with the Yanomamo group in the Amazon basin between Venezuela and Brazil. Tierney accused Chagnon of unethical conduct in the field by violating cultural taboos, introducing weapons, instigating intra-ethnic conflicts, collaborating with a biologist conducting medical research without Yanomamo's informed consent, and even of espionage. Although Tierney's accusations referred to research conducted between the 1960s and the 1980s; they generated a fierce controversy within the anthropological community of the US and Latin America. The American Anthropological Association (AAA) put together a task force to investigate the allegations and published a report that concluded that Chagnon had indeed violated the ethical codes of the discipline (see AAA 2003, Borofski

---

2 Also known as 'culture wars,' such reforms in different academic setting in the US were promoted by progressive scholars who sought to question the study of 'high culture,' which contemplated the study of Western classic literature, by adopting first person testimonial accounts that reflected the life of disenfranchised and excluded people from the new world. See Huizer and Mannheim (1979) and Pratt in Arias (2001).

3 The politics of the 'culture wars' unleashed a controversy around the testimonial affirmation by indigenous activist Rigoberta Menchú and the accusations formulated against her by anthropologist David Stoll. For more insight into this case see Arias (2001).

2005). Members of the AAA working in US academia found themselves revisiting debates from the 1960s about the need to decolonialize the discipline. They found themselves divided along the lines of the objectivity/subjectivity debate. Gross and Plattner (2002), who at the time occupied high positions in research agencies that allocated funds for research projects in the social sciences, advocated so-called detached scholarship over fieldwork and scholarship featuring collaborative relationship between anthropologists and their field subjects. At the core of their argument was the contention that the engagement of the anthropologist with the study community and its members defined as laypersons, can undermine the integrity of research and transform anthropological inquiry into social work, which is supposedly a less scientific undertaking (Gross and Plattner 2002).[4] In 2003 a group of AAA members organized a referendum that asked all AAA members to vote to rescind the Task Force final report allegedly biased against Chagnon. Those who supported the referendum defended that the AAA had no business sanctioning the conduct of an anthropologist as its code of ethics had to be understood as a reference open to reinterpretation. Two-thirds of the voters supported the rescission of the AAA's acceptance of the report (see Gregor and Gross 2004, Lassiter 2005), reconfirming a profound division among US anthropologists regarding political and ethical concerns and their impacts on knowledge production.

This brief historical overview of critiques of positivism in the discipline shows that anthropology, as all other social sciences (and scientific disciplines generally), is not a value-free and detached way of producing knowledge. Ethnography and writing are not simply methods of collecting data but active processes of knowledge production that are situated politically, ethically, and intellectually. In indigenous studies, the increasing participation of indigenous peoples as political actors in national and transnational debates concerning neo-liberalism, market reforms, and other interactions with nation-states and financial institutions means, among other things, that they are not uninformed subjects detached from larger ethical, political, economic, and cultural concerns. Contemporary indigenous movements demand self-determination and protection from incursions upon their sacred and communal lands, as national and multinational economic interests seek access to commodities such as oil, rubber, metals, minerals, and gems in indigenous territories. These contexts call into question the viability of a detached researcher who enters the field site and conducts research regardless of the political implications of such factors. When working with contemporary indigenous communities, anthropologists often find it necessary and inevitable to position themselves in regards of such issues and to reframe their relationship with indigenous actors. I argue that detachment in such contexts cast doubts on the kind of ethnographic knowledge that ethnographers who pretend not to define their positionality in the field can produce (see Calabró and Theodossopoulos in this volume).

---

4   On the objective–subjective debate in anthropology, see *Current Anthropology* 36, 1995, particularly the contribution by D'Andrade and Scheper-Hughes. See also Fabian (2001), Hymes (1972), Roscoe (1995), Salzman (2002) and Sutton (1991).

Critical Reflections and Contributions

Having reaffirmed the situated nature of debates on engagement, I reflect on the specificities of this form on anthropological inquiry in order to identify its strengths and weaknesses. Rather than political propaganda, diverse forms of engagement and collaboration have produced 'excellent' scholarship in anthropology, including the work of anthropologists engaged in activist research and of those who identify themselves as minority members.[5] Feminist, Afro-descent, Latino, queer studies scholars, indigenous scholars, anthropologists from so-called developing countries and activist anthropologists have produced seminal work for the understanding of social processes where gender, race, class tensions, and geopolitics are tightly intertwined and often overlapping.[6] In indigenous studies engaged anthropology, rather than treating indigenous cultures and societies as discrete and exotic objects of anthropological scrutiny, follows a different line of inquiry. Collaborative and engaged practices and methodologies have the potential of decolonizing anthropology by questioning colonialists tropes such as other and otherness, insider and outsider, first world and third world. They focus rather on the conditions and contexts in which indigeneity becomes either a justification for violating territorial or human rights in the name of national and global progress, or for resisting such abuses. My point of departure is the definition of collaboration, or collaborative moment, understood as an epistemological, methodological *and* a political one.

'We' are not alone. In their critiques of the supposedly objective truth of modernist anthropology, feminist, anticolonialist and interpretative scholarship have come to explode the fiction of 'the scientist of culture who works alone' (Lapovsky Kennedy 1995: 26). Various authors addressed this aspect of knowledge production in anthropology by questioning the possibility of accomplishments built on the works of 'lone strangers' (Gottlieb 1995, Rosaldo 1989, Salzman 1994). Such narrative reproduced the mythology of the male adventurer figure and the role he has played in the formation, accumulation, and development of scientific knowledge and spread of ideas in and from the West to the rest of the world. I am thinking of figures such as for example geographer Alexander von Humboldt, whose writings can be considered as a form of proto-anthropology of the South American Andes. These epic accounts extol the journey of adventurous men who defied the unknown for the sake of scientific knowledge, but offer little, if anything about all those who accompanied them on their excursions. In anthropology this 'heroic' tradition was continued by equating the 'lone ethnographer' to the scientist

---

5   I respond here to the discussion on 'excellence' in anthropology in Gross and Plattner (2002).

6   For example, see Abu-Lughod (1991, 1987), Alonso (1995), Aretxaga (1997), Gordon (1998), Gordon, Gurdian and Hale (2003), Hale (2008a, 1997), Jacob-Huey (2002), Limón (1991), Mascia-Lees and Sharpe (2000), Rahier (1999), Rappaport (1990), Torres (1995) and Trix and Sankar (1998).

who faces any hardship in order to prove and test his theories (Rosaldo 1989: 30, 31, Sontag 1966). The figure of the 'lone ethnographer' obscures the inherently interactive process of knowledge production.

Different proposals of collaboration and cooperation in research practice and writing emerged from these critiques. Ever since the 1960s collaboration in different academic settings has highlighted its interdisciplinary (as well as interdepartmental and inter institutional) mode of dialogue and conversation that anthropologists entertain across disciplines with scientists, historians, sociologists, philosophers, cultural studies scholars, students (just to mention a few) in order to elaborate new perspectives on the fast changing cultural and social landscapes.[7] More recently George Marcus has argued that the emergence of new scientific theories about life, genetics, and the environment has made this form of disciplinary crosspollination (not just fieldwork) even more relevant and distinctive for anthropological research (2009, 4–6).

In the field of indigenous studies, however, collaboration or the collaborative moment is understood as both epistemological and highly political since it is premised on the imperative of decolonizing anthropology. In addition to being epistemologically inevitable, as argued above, the collaborative relationship anthropologists establish with indigenous intellectuals and activists pushes the boundaries of knowledge production to other milieus beyond academia. Actors from indigenous forums and organizations, NGOs and other settings are not just interlocutors but initiators of research endeavors whose goals and agendas often differ from research projects crafted within academic settings. The feedback process and the sharing of outcomes produced in such endeavors are reversed: the knowledge and the forms it takes (whether a book, a video, a report, a conference paper, etc.) are tested in non-academic settings and then introduced to academia to fulfill many times the career goals and academic requirements of participating anthropologists. These forms of collaborations are unique in foregrounding the political goals that the anthropological knowledge is intended to accomplish (train activists, inform policies and political demands, foster change in communities, facilitate networking, etc.). The richness of such anthropological engagement lays, among other things, in the production of different types of texts each of which follows a specific set of criteria that operate in different scales (localized, national, transnational) and sites (community, organizations, state institutions, streets, academia). Each of these texts has its own internal coherence, logic and audience but it also is a piece in a multi-textual form of inquiry and hermeneutics. I am thinking here of a collaborative research effort with indigenous communities and organizations that can lead to the elaboration of different products (for example a

---

7   In the US Marcus (2009) has reformulated collaboration as an epistemological and methodological dimension that followed the reflexive turn in US anthropological debates. In the US academic debates of the collaboration of anthropologists and cultural studies scholars, among others, led to the post-modern critique epitomized by *Writing Culture.*

video, a bilingual text, and an academic paper) all of which offer a very different and yet interrelated perspective on the same research topic.

During my first fieldwork in Ecuador I did research in collaboration with a grassroots indigenous organization, the Inca Atahualpa in the parish of Tixán, in the Chimborazo province. The work we did together for two years produced a small bilingual text that was supposed to be used in local schools on the history of the Inca Atahualpa, several videos on local indigenous traditions made by one of the Inca's leaders, proposals for a development project to NGOs, and more recently my scholarly book (see Cervone 2012) and several articles that were published in Ecuador and in the US. Each of these products followed different parameters, were meant to reach different types of audiences, presented a different perspective on the political process it represented, and fulfilled different goals. Each of them therefore presented a different point of view and a different vision of the political process experienced by the Quichuas of Tixán, offering a good sample of the multi-sided nature of their experience.

The collaboration between engaged anthropology and indigenous studies can produce a multi-textual hermeneutics, where the multi-textuality is a way of interpreting and representing the complex and multifaceted aspects of a given situation. It also represents and embodies the different and at times contradictory positioning of the actors involved as an inherent element of the collaborative process itself.[8] The complexity that can be revealed by the intertwining of different texts can provide the ethnographic 'grounding' that Deepa Reddy sees as important to make sense of a field of inquiry that is otherwise disjointed and disconnected (Reddy 2009: 95). More reflection is needed on the intertwining of such multi-textual forms of knowledge production and their contributions to the debate on epistemology, ethics and methodology in anthropology. I argue that such multi-textuality can address the complexity of the 'global' world, understood as a process of cultural, physical, socio-economic and political intertwining in which the mapping of inequality is simultaneously ever present and shifting (Inda and Rosaldo 2007).

However, collaborative methodologies have their own challenges and potential weaknesses (Lamphere 2003, Heckler and Langton in this volume). The political nature of such collaboration can present possible ethical, methodological and epistemological dilemmas. In his study of racial ambivalence among Ladinos in Guatemala, Charles Hale (2008b) reveals the challenges that he faced as an activist anthropologist allied with the Mayan indigenous movement when embarking on fieldwork practice with the supposed dominants, the Ladinos. For

---

8   A controversial case of indigenous justice during my fieldwork in Tixán is good example of this complexity. In that case corporal punishment was made redundant due to local organizations competing for political legitimacy, and that generated a passionate debate among indigenous activists and their supporters on the potential contradictions of such local 'ancestral' practices and the principles stated by human rights conventions that the indigenous movements refer to in their struggles (see chapter 5 in Cervone 2012).

his study of racial politics, Hale conducted fieldwork with Ladinos occupying a wide political spectrum with respect to Mayan activism, ranging from racial prejudice to a declared commitment to cultural equality. Hale suggests that his overt positioning as a Maya supporter and the omission from his analysis of those voices who did not give permission to be disclosed were enough to safeguard both his ethical and activist stand. Yet, even if anonymous, the unwilling voices of Ladinos are represented in Hale's account, revealing that political commitment in circumstances like those he describes may clash with professional ethics. Different ethical standards seem to be at play when research is carried out with those who retain power and privileges in society. The proposal of Rodolfo Stavenhagen (1971) to decolonize anthropology by doing research on the 'powerful' appears more complex than expected.

Another tension inherent in engaged anthropology relates to the actual collaboration in the field and the challenges posed by the motivations of the researchers that emerge in their manifest positionality. In the debate on reflexivity in anthropology some anthropologists have addressed the risk of falling into narcissist and ethnocentric self-referentiality and navel-gazing (Fabian 2001, Friedman 1994, Salzman 2002, Sangren 1988). In engaged anthropological practice, the challenge is to avoid the projection of the researcher's ideals and hopes onto the political processes under analysis. In other words, general principles assuming of anti-colonialism and anti-oppression can be a risky basis for cooperation if they lead the anthropologists and their research partners falsely to assume that they share the same understanding of the path to be followed in the pursuit of justice. Regarding indigenous movements in the Latin America, how should we understand their struggle for self-determination in multicultural societies? Collaborative studies of such struggles highlight complex and often contradictory understandings of change, democracy, and even of human rights (Cervone 2012, De La Peña 2002, Gow and Rappaport 2002, Hale 1994, Rappaport 2005).

In this context, what happens when such possible discrepancies and tensions make collaboration in the field difficult or even conflictive? In other words, what happens when collaboration shows its multifaceted and rhizomic nature, when it becomes 'too weak a word to describe the entanglements that are by now thoroughly commonplace in cultural anthropology: entanglements of complicity, responsibility, mutual orientation, suspicion and paranoia, commitment and intimate involvement, credit and authority, and the production of reliable knowledge for partially articulated goals set by organizations, institutions, universities, corporations, and governments' (Kelty 2009: 205)? Paraphrasing Kelty: is collaboration a 'too feel-good or too friendly term for the commitment, fights and compromises' that all the actors involved in these relationships experience in the pursuit of their goals (ibid.)? In relation to indigenous studies I believe engagement as a form of anthropological inquiry focuses on the complexity of global forms of discrimination and offers the opportunity of reversing power relationship in anthropological practice. In other words, engagement is an anthropological approach that is adequate to 'the problems of our time' by revealing the multiplicity of actors, forces and dimensions (local, national, and transnational) that figure in the definition of ethnic identities and influence the identification of indigenous cultures (Wolf in Berreman 1968: 395). Yet, like other forms of anthropological inquiry, engaged anthropology has no formulaic answers to provide to its dilemmas other than nurturing self-critique and being suspicious of any sense of accomplishment and 'feel-goodness.'

Finally, a perhaps more abstract question about anthropological epistemology. If we agree, as I think many of us do, on the possibility of anthropology in the plural, to what extent is it possible to produce an alternative anthropological knowledge that is not linked to the theoretical debates and purposes that historically have oriented this discipline? For example how can we approach the study of identity whatever our ethnic and cultural background, without referring, either to affirm or to refute, theoretical debates around that topic (essentialism, instrumentalism, constructivism and post-constructivism)? Even when engagement in debates with indigenous scholars and activists challenges such theoretical approaches, it does not challenge one of the major epistemological premises that historically has defined the discipline. One such premise posits that anthropology involves cultural translation, initially needed to supposedly render intelligible to colonial regimes what was not intelligible about their subjects. The relationships of power implicated in such relations have changed but the basic premise of translating and making intelligible often remains. Whether to control, or to correct a misinterpretation, or to denounce an injustice, anthropology works an act of translation and of interpretation that occurs within the parameters of a Euro-centered theoretical and epistemological narrative. Are there, then, only different possible anthropological translations rather than different anthropologies? What would anthropology be without an act of translation? If translation is anthropology's straight jacket, how can we decolonize such translations?

Decolonizing Methodologies: An Open Field

In the wake of Said's *Orientalism* the analyses of socio-cultural imaginaries, which accompanied the many remapping of geopolitical power structures, have identified the major sources of inequalities in early modern expansions of colonial and imperial regimes, and more recently in the global expansions of late capitalism. Although constantly challenged and redefined, such inequalities and their sources endure and are even reinforced by contemporary global processes (Harvey 2005). Scholars such Walter Mignolo (2000), and Aníbal Quijano (2000) have analysed such questions by focusing on the historical process from which the idea of the West emerged to produce and perpetuate what they define as the 'coloniality of power'; that is, a conceptual and territorial system of inequality informing relationships between people, places and ideas. Such a system is what still sustains, according to the authors, the expansion of the global capitalist economy albeit with significant emerging shifts in geopolitics. In the realm of indigenous studies the debate about the need to decolonize anthropological knowledge has focused on what Trouillot calls the 'savage slot' which defines the inferiors' slot, subsequently redefined and reproduced in different historical and socio-political contexts.

In Latin America, the process of politicization of ethnic identity and the anti-discrimination struggle of indigenous movements has led to a shift in anthropological research interests and methodologies (Rappaport 1994, 17). Since the early 1990s, studies of indigeneity have highlighted the historical and political nature of ethnic identity formation among indigenous people.[9] Urban and Sherzer (1991) paved the way for new studies of indigeneity, which focus on the complex and ever-changing nature of a political process that involves both indigenous actors and nation states. Contemporary studies of indigenous peoples examine how they engage with economic policies and social changes at the national and transnational levels to negotiate their position of subordination vis-à-vis states and nonindigenous groups.[10]

This shift, which is not restricted to Latin America, represents a response to the epistemological necessity I mentioned above. It also reflects the collaborative turn in as much as it underscores the presence and participation of previously silenced voices in the production of anthropological knowledge. Increasing numbers of indigenous scholars have been concerned with producing their own interpretations about their societies by focusing on the revitalization of values and cultural systems which had been denigrated and misrepresented by decades of colonial and postcolonial governance (see CONAIE 1996, Harry 2009, LaDuke 1999, 2005, Smith 2012 to name a few). The fields of education and pedagogy, and more recently the field of environmental justice, have been the major arenas in which the voice of indigenous scholars and activists has become louder and more assertive. One pioneer text in this respect is Lynda Smith's *Decolonizing Methodologies* which has paved the way for many indigenous scholars and anthropologists to start thinking about their own learning processes and breaking with pedagogical criteria imposed by the dominant non-native societies. The decolonizing impact of such interventions is twofold: on the one hand it led to the recognition in academia of the presence of indigenous scholars and to the acknowledgement of the validity and vitality of their scholarship.[11] On the other hand, it has provided indigenous communities and organizations with new tools to elaborate their own alternative models of education, as well as negotiate their active participation in the design and implementation of policies and research.

---

9   Abercrombie (1991a, 1991b), Barre (1983), Botasso-Gnerre (1989), Cervone (2012), Maria Elena García (2005), Gustafson (2009), Mattiace (2003), Pallares (2002), Postero (2007), Rappaport (1994, 2005), Sawyer (2004) and Warren (1998), among others.

10   Colloredo-Mansfeld (1999), Kearney (1996), Lagos (1994), Meisch (2002), Nash (2001), Orlove (2002), Weismantel (1988) and Warren and Jackson (2003).

11   However, the decolonization of anthropology and academia is to be understood as a work in progress since, as Karen Brodkin, Sandra Morgen and Janis Hutchinson (2011) argue in the case of the US, academia remains mostly a 'white public space.'

The collaborative moment offers great opportunities today for experimentation with new teaching methodologies, exposing students in different academic settings to new perspectives and points of view concerning contemporary indigenous societies. In the context from which this book emerged, a group of anthropologists and indigenous scholars elaborated a proposal for a collaborative and co-taught course on Indigenous Agency and Innovations to be offered in different institutions, according to which various scholars and activists would offer a lecture to be recorded and delivered via podcast or via live conference call. I taught the course in the fall of 2010 even though the proposed methodology did not work out the way it was originally planned.[12] I had some institutional support and a small grant to fund three guest speakers to talk on topics that ranged from biocolonialism, to food security and education in different indigenous communities.

For many of the students it was the first time they had met and listened to an indigenous speaker and scholar. Their direct voice as activists who struggle against encroachment on their lands, exclusion and racism in their everyday life had a profound impact on students' capacity to relate to those issues. Their interventions generated interesting debates on what it means to be Native American and indigenous today. The rejection of genetic sampling in indigenous territory presented by Native American activist Debra Harry, for example, sparked a lively discussion on cultural diversity, race and power relations in contexts in which Native Americans and other indigenous groups have been treated as second class citizens ever since the formation of modern nations-states. The direct, personal and human experience of the speakers made their struggle come alive in the classroom, with mixed reactions from the students. Some, when pushed out of their comfort zone, could not accept that their own model of life and value system were not the same one embraced by the speakers, and objected to the validity of their claims as being exclusionary or even going against the interest of humanity at large. If genetic sampling can foster medical research, some of them argued, why are Native Americans hindering that process? Some others had a sort of epiphany and clearly grasped the eminently political nature of indigenous claims, whether they touched upon genetic sampling or food security.

This experience confirmed my belief that such collaborative pedagogical modalities, made possible by new technologies, can represent another step towards a decolonized anthropological pedagogy and academia.

Decolonization from Within

Decolonizing is a complex inner process that involves political engagement on several fronts. Many decolonizing efforts and practices focus on the relationship

---

12   Difficulty in the synchronizing of academic calendars and resources made the co-teaching particularly challenging. Yet, I believe this type of proposals should be revisited and implemented as viable forms of collaborative methodologies.

between indigenous peoples vis-á-vis the state and the non-indigenous citizens as a way to support anti-discriminatory policies and struggles. I argue that equal attention needs to be put on the perverse modalities with which racism, discrimination, and exclusion still affects (and in some cases alter) relationships and interactions *within* indigenous societies (Harrison 1997). Intellectuals such as Frank Fanon (2005, 2008) and W.B. De Bois (1994), among others, have cogently highlighted the pernicious and devastating impacts of racism on those who have been targeted as inferior. The effects of decades, and centuries, of humiliation and violence have led to a process of internalization of such annihilating ideas and practices, which perpetuated discrimination internally. The outcomes of these internal forms of violence vary according to the specificity of the contexts in which they take place. I turn to the case of violence against indigenous women in Ecuador, on which I have worked for many years, as an example that calls for the need of what I define decolonization from within.

Debates on violence against indigenous women in their own communities have started to emerge in Ecuador in the late 1990s. Ever since then many indigenous women have organized to fight against such forms of violence but they face insurmountable barriers in their quest for equality and a life free of violence within their own communities. Their struggle soon turned into a highly controversial issue within indigenous communities and organizations (Cervone 1998). The acknowledgement of those negative and problematic practices is seen as dangerous for the integrity of a movement that has made of the politicization of ethnic identity a powerful tool to fight against discrimination. How can anthropology contribute to social change in such a case, and participate in the decolonization of decades, even centuries, of racist violent practices that permeate relationships of power within indigenous communities?

These questions motivated a group of feminist researchers and indigenous women activists to combine their work and interests on issues of gender equality in a comparative and collaborative research project on indigenous women's rights and indigenous justice in Latin America.[13] This collaboration of anthropologists and indigenous women's activists aims to shade a new light on the root-causes of gender violence within indigenous communities. As a member of that group I examine my participation in the research project to discuss the potential of collaborative methodologies to foster a process of internal decolonizing practice.

Recent scholarship on gender violence in the Ecuadorian Andes highlights the ambiguities and contradictions inherent in contemporary discourses on indigenous justice and gender equality.[14] Studies examine the inadequacies that indigenous justice systems present when faced with cases of domestic violence and abuses against women. Such forms of violence are attributed to the sexisms of indigenous males and to the condition of triple discrimination that women experience for being poor, women and indigenous. Although such critical insights are important

---

13   The project is run by CIESAS, Mexico and coordinated by Rachel Sieder.
14   See Picq (2012), Pequeño (2009).

contributions to the analysis of gender violence in indigenous communities, existing scholarship on racialized societies as well as the complex political context in which such issues are debated require a more nuanced interpretation. My collaboration with my research partner and indigenous activist Cristina Cucurí was pivotal in exploring the problem from a different perspective.[15] Cristina and I decided to delve more into the root-causes of such violence by building on a question that Cristina's experience as an activist brought to the table: Why do indigenous men treat their women so badly? We both decided that it was paramount to our research to also collect the point of view of men.

In a series of interviews and life histories with Quichua indigenous women and men from the highland province of Chimborazo we stumbled upon the legacy of patterns of violence and abuse that had become part of people's memory and daily life. Many people we interviewed (elderly and younger women and men) connected the different forms of violence that women, and to certain degrees children, suffer in their families and communities to patterns of abuse and rejection against indigenous people that are still common in Ecuadorian society. Those patterns are locally entangled with a history of power relations dating back to the beginning of the 1900s when indigenous people in the highlands were living in a state of de facto slavery under powerful landowners. In that context violence, rape and corporal punishment functioned as mechanisms to impose obedience and control and to reproduce racialized hierarchies both locally and nationally. The modernization of the agrarian structure of the country that came with the agrarian reforms since the 1960s and the economic policies that affected the rural economy ever since then did not eliminate patterns of exclusion and prejudice against indigenous citizens (see Becker 2008, Bretón 2012, Cervone 2012,). Many men we interviewed denounced the different forms of discrimination they still faced in their everyday life in schools, or every time they had to apply for a loan, entered a public office, or selling their produce in the market place. How to make sense of the voices we had collected in our interviews and the macro context of policies, politics and changes that had seen the affirmation of the indigenous movement and its anti-discrimination struggle while still reproducing practices of racial exclusion and prejudice?

By combining scholarship and the political experience of my research partner we worked to interpret such complexities in a way that could be theoretically sound and politically meaningful. Fanon's theories on the internalization of racism prove especially helpful to understand our case. We argued that the violence women experience today cannot be understood just by foregrounding sexist practices but needs to be connected to the history and continuity of racist practices and prejudice.

---

15 Cristina Cucurí is Quichua from Chimborazo and the president and co-founder of the REDCH, Network of Indigenous Women's Organizations of Chimborazo. She also participated in the national elections for congress in 2009 with the Pachakutik party. The REDCH political struggle for women's rights was pivotal in having indigenous women's right to political participation recognized in the new constitution of 2008.

The recurrence of physical and psychological violence, of discrimination and even of sexual abuse that emerged in our interviews attests that violence has made its way within indigenous societies as a legitimate way to sanction obedience and exert control into marital relationships and children's upbringing.

We also agreed with the other studies on the paradox of the recent multicultural turn. While the constitutional recognition of indigenous cultural diversity reinforces the perception of cohesion and harmony as positive values of indigenous cultures, such perception makes the fight against inner forms of violence and exclusion more difficult to sustain.[16] According to Cristina, understanding violence against women in a broader framework of racist violence can also help to incorporate women's political agenda within a larger and more overarching anti-racist struggle undertaken by the indigenous movement.

It is important for decolonizing methodologies and pedagogies, therefore, to reflect on the complex implications of racism and discrimination, and on the process of internalization on many levels both external (often undetected in non-indigenous advocacy) as well as internal to indigenous societies. In Ecuador this has proved to be a very difficult task due to the level of politicization of the indigenous movement over the decades. Leadership within the movement by indigenous men is the result of a national political culture that has delegitimized and rendered invisible female political participation. This leadership pattern makes it very difficult for women to implement anti-violence agendas in their own communities and to have indigenous authorities participate to control it.[17] Engendered patterns of indigenous power and leadership end up reproducing the same patterns of discrimination and exclusion imposed by the dominant white society. An anti-racist framework can offer a window of opportunity that could lead to new forms of justice within indigenous communities aiming to eradicate any form of violence. It is, Cristina says, a challenging path but not an impossible one.

In such contexts, collaborative research is more complex and multifaceted than anticipated, rendering the positioning of the researcher more problematic. Internal contradictions add another layer of complexity to the researchers' support of indigenous movements' anti-discrimination struggles. How do we, scholars and activists alike, account for the controversial knowledge that we believe can help social change? How do we account for the personal safety of our research partners who are involved in such complex realms? Again, I believe that anthropology can contribute by challenging assumptions of 'feel-goodness' in collaborative methodologies on the one hand, and by producing critical knowledge that is skeptical of easy rendering of political engagements and solidarity.

---

16   The new Ecuadorian constitution of 2008 sanctions the right to cultural and ethnic diversity of indigenous peoples and their right to self-determination.

17   Some cases of domestic violence involve male leaders. See case in Picq (2012).

Concluding Remarks

Engagement, like any other form of anthropological practice, is not an unproblematic, universally valid, or flawless form of inquiry. It is imperative that debates about the future of collaboration are fostered and supported together with the constant search for new epistemologies and methodologies. Several points for discussion arise from the examination of the validity and limitations of the collaborative moment. For instance, the situated nature of this moment: How might such engagement and forms of collaboration evolve and change once indigenous actors, such as the Quichuas in Ecuador, become involved in new forms of governance and politics in their own countries? How would the redefinition of their position in society impact on collaboration, cooperation, and positionality? What kinds of ethical, methodological and epistemological concerns should be considered when the agendas and interests of different indigenous actors and activists diverge? Like other social science disciplines, anthropology faces new challenges posed by the increasing complexity inherent in globalized societies, where distinctions between powerless and powerful are often blurred and where systems of stratification combine class, ethnicity, race, gender, and geography in varying and multifaceted ways. How do all these old and new social identities and structures impact on questions of positionality, epistemology, and ethics? The anthropological profession today faces such open-ended questions. They are best addressed by taking into consideration the different perspectives, experiences, and points of views of the many voices involved in the collaborative moment as well as the multitextual nature (that I highlighted earlier) of the knowledge produced in collaboration.

References

Abercrombie, T. 1991a. To be Indian, to be Bolivian: 'Ethnic' and 'National' Discourses of Identity, in *Nation-States and Indians in Latin America*, edited by G. Urban and J. Sherzer. Austin, TX: University of Texas Press, 95–130.

—— 1991b. Articulación Doble y Etnogénesis, in *Reproducción y Transformación de las Sociedades Andinas* vol. 1, edited by Segundo Moreno and Frank Salomon. Quito: AbyaYala, 197–212.

Abu-Lughod, L. 1987. *Veiled Sentiments: Honor and Poetry in a Bedouin Society*. Cairo: American University in Cairo Press.

—— 1991. Writing Against Culture, in *Recapturing Anthropology Working in the Present*, edited by R. Fox. Santa Fe, NM: School of American Research Press, 137–62.

Alonso, A. 1995. *Thread of Blood: Colonialism, Revolution, and Gender on Mexico's Northern Frontier*. Tucson, AZ: University of Arizona Press.

American Anthropological Association 2003. *El Dorado Task Force Final Report* 1(2.3). Mimeo.

Aretxaga, B. 1997. *Shattering Silence: Women, Nationalism and Political Subjectivity in Northern Ireland*. Princeton, NJ: University of Princeton Press.

Arias, A. (ed.) 2001. *The Rigoberta Menchú Controversy*. Minneapolis, MN: University of Minnesota Press.

Barre, M.C. 1983. *Ideologías Indigenistas y Movimientos Indios*. Spain: Siglo XXI.

Becker, M. 2008 *Indians and Leftists in the Making of Ecuador's Modern Indigenous Movements*. Durham, NC: Duke University Press.

Berreman, G.D., Gjessing, G. and Gough, K. 1968. Social Responsibilities Symposium. *Current Anthropology* 9(5), 391–436.

Borofsky, R. (ed.) 2005. *Yanomami: The Fierce Controversy and What We Can Learn from It*. Berkeley, CA: University of California Press.

Botasso, J. and Gnerre, M. (eds) 1989. *Del Indigenismo a las Organizaciónes Indigenas*. Quito: AbyaYala.

Bretón, V. 2012. *Toacazo. En los Andes Equinocciales tras la reforma agraria*. Quito: FLACSO—Abya Yala.

Brodkin, K., Morgen, S. and Hutchinson, J. 2011. Anthropology as White Public Space? *American Anthropologist*, New Series, 113(4), 545–56.

Cervone, E. 2002. Engendering Leadership. Indigenous Female Leaders in the Ecuadorian Andes, in *Gender's Place: Feminist Anthropologies of Latin America*, edited by L. Frazer, J. Hurtig and R. Montoya. New York: Palgrave, 179–96.

—— 2007. Building Engagement: Ethnography and Indigenous Communities Today. *Transforming Anthropology* 15(2). John Wiley & Sons, 97–110.

—— 2012. *Long Live Atahualpa. Indigenous Politics, Justice, and Democracy in the Northern Andes*. Durham: Duke University Press.

Colloredo-Mansfeld, R. 1999. *Native Leisure Class. Consumption and Culture Creativity in the Andes*. Chicago: University of Chicago Press.

CONAIE 1989. *Las Nacionalidades Indígenas en el Ecuador. Nuestro Proceso Organizativo*. Quito: Abya Yala- Tincui.

D'Andrade, R. 1995. Moral Models in Anthropology. *Current Anthropology* 36(3), 399–408.

De la Peña, G. 2002. Social Citizenship, Ethnic Minority Demands, Human Rights and Neoliberal Paradoxes: A Case Study in Western Mexico, in *Multiculturalism in Latin America: Indigenous Rights, Diversity, and Democracy*, edited by R. Sieder. New York: Palgrave, 129–56.

Du Bois, W.E.B. 1994. *The Soul of Black Folk*. New York: Dover Publishers.

Fabian, J. 2001. *Anthropology with an Attitude*. Stanford, CA: Stanford University Press.

Fanon, F. 2005. *The Wretched of the Earth*. New York: Grove Press.

—— 2008. *Black Skin White Masks*. New York: Grove Press.

Friedman, J. 1994. The Past in the Future: History and the Politics of Identity. *American Anthropologist* 94(4), 837–59.

García, M.E. 2005. *Making Indigenous Citizens: Identity, Development and Multicultural Activism in Peru*. Stanford, CA: Stanford University Press.

Gordon, E.T. 1998. *Disparate Diasporas: Identity and Politics in an African Nicaraguan Community*. Austin, TX: University of Texas Press.

Gordon, E.T., Gurdian, G. and Hale, C. 2003. Rights, Resources, and the Social Memory of Struggle: Reflections on a Study of Indigenous and Black Community Land Rights on Nicaragua's Atlantic Coast. *Human Organization*, 62(4), 369–81.

Gottlieb, A. 1995. Beyond the Lonely Anthropologist: Collaboration in Research and Writing. *American Anthropologist* 97(1), 21–6.

Gow, D. and Rappaport, J. 2002. The Indigenous Public Voice: The Multiple Idioms of Modernity in Native Cauca, in *Indigenous Movements, Self-Representation, and the State in Latin American*, edited by K. Warren and J. Jackson. Austin, TX: University of Texas Press, 47–80.

Gregor, T. and Gross, D. 2004. Guilt by Association: The Culture of Accusation and the American Anthropological Association's Investigation of Darkness in El Dorado. *American Anthropology* 106(4), 687–98.

Gross, D. and Plattner, S. 2002. Anthropology as Social Work: Collaborative Models of Anthropological Research. *Anthropology Newsletter* (American Anthropological Association) 43(8), 4.

Gustafson, B. 2009. *New Languages of the State: Indigenous Resurgence and the Politics of Knowledge in Bolivia*. Durham, NC: Duke University Press.

Hale, C. 1994. *Resistance and Contradiction: Miskitu Indians and the Nicaraguan State, 1894–1987*. Stanford, CA: Stanford University Press.

—— 1997. Cultural Politics of Identity in Latin America. *Annual Review of Anthropology* 26, 567–90.

—— 2008a. *More Than an Indian. Racial Ambivalence and Neoliberal Multiculturalism in Guatemala*. Sante Fe, NM: School of American Research Press.

—— 2008b. Introduction, in *Engaging Contradictions. Theory, Politics and Methods of Activist Scholarship*, edited by C. Hale. Berkeley, CA: University of California Press, 1–28.

Haraway, D. 1988. Situated Knowledges: The Science Question in Feminism and the Privilege of Partial Perspective. *Feminist Studies* 14(4), 575–99.

Harrison, F.V. (ed.) 1997 [1991]. *Decolonizing Anthropology: Moving Further toward an Anthropology of Liberation*. Washington, DC: American Anthropological Association.

Harry, D. 2009. *Indigenous Peoples and Gene Disputes. Chicago-Kent Law Review* 84, 1–47.

Harvey, D. 2005. *A Brief History of Neoliberalism*. Oxford, UK: Oxford University Press.

Huizer, G. and Mannheim, B. (eds) 1979. *The Politics of Anthropology: From Colonial-ism and Sexism toward a View from Below*. The Hague: Mouton Publishers.

Hymes, D. (ed.) 1972. *Reinventing Anthropology*. New York: Vintage.

Inda, X. and Rosaldo, R. (eds) 2007. *Anthropology of Globalization: A Reader*. Blackwell Publishing.

IWGIA 1971. *Declaration of Barbados*. IWGIA Documents. Denmark.

Jacobs-Huey, L. 2002. The Natives Are Gazing and Talking Back: Reviewing the Problematics of Positionality, Voice, and Accountability among 'Native' Anthropologists. *American Anthropologist* 104(3), 719–804.

Kearney, M. 1996. *Reconceptualizing the Peasantry Anthropology in Global Perspective*. Boulder, CO: Westview Press.

Kelty, C. 2009. Collaboration, Coordination, and Composition: Fieldwork after the Internet, in *Fieldwork is not What it Used to Be*, edited by G. Marcus and J. Faubion. Ithaca: Cornell University Press, 184–206.

LaDuke, W. 1999. *All Our Relations: Native Struggles for Land and Life*. Cambridge, MA: South End Press.

—— 2005. *Recovering the Sacred: The Power of Naming and Claiming*. Cambridge, MA: South End Press.

Lagos, M. 1994. *Autonomy and Power: The Dynamics of Class and Culture in Rural Bolivia*. Philadelphia, PA: University of Pennsylvania Press.

Lamphere, L. 2003. The Perils and Prospects for an Engaged Anthropology. A View from the United States. *Social Anthropology* 11(2), 153–68.

Lapovsky Kennedy, E. 1995. In Pursuit of Connection: Reflections on Collaborative Work. *American Anthropologist* 97(1), 26–33.

Lassiter, E. 2005. Collaborative Ethnography and Public Anthropology. *Current Anthropology* 46(1), 83–106.

Limón, J. 1991. Representation, Ethnicity and the Precursory of Ethnography. Notes of a Native Anthropologist, in *Recapturing Anthropology Working in the Present*, edited by R. Fox. Santa Fe, NM: School of American Research Press, 115–35.

Marcus, G.E. 1986. Contemporary Problems of Ethnography in the Modern World System, in *Writing Cultures*, edited by G. Marcus and J. Clifford. Berkeley, CA: University of California Press, 165–93.

Marcus, G.E. and Clifford, J. (eds) 1986. *Writing Cultures*. Berkeley, CA: University of California Press.

Marcus, G.E. and Faubion, J. (eds) 2009. *Fieldwork is Not What it Used to Be*. Ithaca: Cornell University Press.

Mascia-Lees, F. and Sharpe, P. 2000. *Taking a Stand in a Postfeminist World*. Albany, NY: SUNY Press.

Mattiace, S. 2003. *To See with Two Eyes: Peasant Activism and Indian Autonomy in Chiapas, Mexico*. Albuquerque, NM: University of New Mexico Press.

Mead, M.1943. The Role of the Scientist in Society, in *Anthropology: A Human Science. Selected Papers, 1939–1960*, edited by M. Mead. Princeton, NJ: D. Van Nostrand Company, 85–91.

Meisch, L. 2002. *Andean Entrepreneurs. Otavalo Merchants and Musician in the Global Arena*. Austin, TX: University of Texas Press.

Mignolo, W. 2000. *Local Histories/Global Designs: Coloniality, Subaltern Knowledges, and Border Thinking. Princeton Studies in Culture/Power/ History*. Princeton, NJ: Princeton University Press.

Morin, F. and Saladin d'Anglure, B. 1997. Ethnicity as a Political Tool for Indigenous Peoples, in *The Politics of Ethnic Consciousness*, edited by C. Govers and H. Vermeulen. London: St Martin's Press, 157–93.

Muratorio, B. 1991. *The Life and Time of Grandfather Alonso*. New Brunswick, NJ: Rutgers University Press.

Nash, J. 2001. *Mayan Visions. The Quest for Autonomy in an Age of Globalization*. New York: Routledge.

Orlove, B. 2002. *Lines in the Water. Nature and Culture at Lake Titicaca*. Berkeley, CA: University of California Press.

Pallares, A. 2002. *From Peasant Struggles to Indian Resistance: The Ecuadorian Andes in the Late Twentieth Century*. Norman, OK: University of Oklahoma Press.

Pequeño, A. 2009. *Participación y políticas de mujeres indígenas en América Latina*. Quito: FLACSO, Ecuador and Ministerio de Cultura del Ecuador.

Picq, M. 2009. Between the Dock and a Hard Place: Hazards and Opportunities of Legal Pluralism for Indigenous Women in Ecuador. *Latin American Politic and Society* 54(2), 1–33.

Postero, N. 2007. *Now We are Citizens: Indigenous Politics in Postmulticultural Bolivia*. Stanford, CA: Stanford University Press.

Quijano, A. 2000. Coloniality of Power, Eurocentrism and Latin America. *Nepantla: Views from the South* 1(3), 533–80.

Rahier, J. (ed.) 1999. *Representations of Blackness and the Performance of Identities*. Westport, CT: Bergin and Garvey.

Rappaport, J. 1990. *The Politics of Memory: Native Historical Interpretation in the Colombian Andes*. Durham, NC: Duke University Press.

—— 1994. *Cumbe Reborn: An Andean Ethnography of History*. Chicago, IL: University of Chicago Press.

—— 2005. *Intercultural Utopias: Public Intellectuals, Cultural Experimentation, and Ethnic Pluralism in Colombia*. Durham, NC: Duke University Press.

Reddy, D. 2009. Caught! The Predicaments of Ethnography in Collabroation, in *Fieldwork is Not What it Used to Be*, edited by G. Marcus and J. Faubion. Ithaca: Cornell University Press, 89–112.

Rosaldo, R. 1989. *Culture and Truth. The Remaking of Social Analysis*. Boston, MA: Beacon Press.

Roscoe, P. 1995. The Perils of Positivism in Cultural Anthropology. *American Anthropologist* 9(3), 492–504.

Salzman, P. 1994. The Lone Stranger in the Heart of Darkness, in *Assessing Cultural Anthropology*, edited by R. Borofsky. Cambridge, MA: McGraw-Hill, 29–39.

—— 2002. On Reflexivity. *American Anthropologist* 104(3), 805–13.

Sangren, S. 1988. Rhetoric and the Authority of Ethnography: 'Postmodernism' and the Social Representation of Texts. *Current Anthropology* 29(3), 405–35.

Sanjek, R. 2004. Going Public: Responsibilities and Strategies in the Aftermath of Ethnography. *Human Organization* 63(4), 444–56.

Sawyer, S. 2004. *Crude Chronicles. Indigenous Politics, Multinational Oil, and Neoliberalism in Ecuador*. Durham, NC: Duke University Press.

Scheper-Hughes, N. 1995. The Primacy of the Ethical. Propositions for a Militant Anthropology. *Current Anthropology* 36(3), 409–20.

Slocum, S. 1975. Woman the Gatherer: Male Bias in Anthropology, in *Toward an Anthropology of Women*, edited by R. Reiter. New York: Monthly Review Press, 36–50.

Smith Tuhiwai, L. 2012. *Decolonizing Methodologies: Research and Indigenous Peoples*. Second Edition. London, UK: Zed Books.

Sontag, S. 1966 [1963]. The Anthropologist as Hero, in *Against Interpretation*, edited by S. Sontag. New York: Farrar, Strauss and Giroux, 69–81.

Stavenhagen, R. 1971. Decolonizing Applied Social Sciences. *Human Organization* 30(4), 333–46.

Stoll, D. 1999. *Rigoberta Menchú and the Story of All Poor Guatemalans*. Boulder, CO: Westview Press.

Tierney, P. 2000. *Darkness in El Dorado*. New York: Norton.

Torres, A. 1995. *Blackness, Ethnicity and Cultural Transformation in Southern Puerto Rico*. PhD dissertation, Department of Anthropology, University of Illinois at Urbana-Champaign.

Trix, F. and Sankar, A. 1998. Women's Voices and Experiences of the Hill–Thomas Hearing. *American Anthropologist* 100(1), 32–40.

Trouillot, M.R. 1991. Anthropology and the Savage Slot. The Poetics and Politics of Otherness, in *Recapturing Anthropology Working in the Present*, edited by R. Fox. Santa Fe, NM: School of American Research Press, 17–44.

Urban, G. and Sherzer, J. 1991. *Nation-states and Indians in Latin America*. Austin, TX: University of Texas Press.

Warren, K. 1998. *Indigenous Movements and Their Critics. Pan Maya Activism in Guatemala*. Princeton, NJ: Princeton University Press.

Warren, K. and Jackson, J. (eds) 2003. *Indigenous Movements, Self-Representation and the State in Latin America*. Austin, TX: University of Texas Press.

Weismantel, M. 1988. *Food, Gender, and Poverty in the Ecuadorian Andes*. Philadelphia, PA: University of Pennsylvania Press.

Chapter 6

Urban Amerindians and Advocacy: Toward a Politically Engaged Anthropology Representing Urban Amerindigeneities in Manaus, Brazil

J.P. Linstroth

In Brazil many anthropologists are activists and advocates for indigenous rights and causes. This praxis is quite distinguishable from North American and British anthropological traditions. As Ramos (2012: 490) explains: 'generations of Brazilian anthropologists ... believe that academic work and political engagement go hand in hand'. Such activism and advocacy for many Brazilian anthropologists 'reflects on the character of his [or her] research, his [or her] choice of topics, of theoretical approaches, fieldwork strategies, and ethnographic writings' (Ramos 1990: 455; see also, Ramos 2000, 2003). The Department of Anthropology at the Universidade Federal do Amazonas (UFAM) in Manaus, Brazil – where I was a visiting professor in 2009 – was no exception to this norm of political involvement with indigenes throughout Brazil.[1] The Brazilian anthropologists at UFAM were always inviting various Amerindian groups to meetings, conferences, and seminars, which at times included the media. UFAM anthropologists many times were also instrumental in negotiating with government organizations, dealing with non-governmental organizations (NGOs) and in addressing concerns such as health and education as well as being involved in legal issues on behalf of indigenous peoples.

It was in this atmosphere that I was situated as an anthropologist where engaging on multiple levels with Brazilian Indians was the ubiquitous standard. I was therefore to be an engaged anthropologist by accident because of these circumstances and because of the nature of local indigenous politics. The expectation was for me to collaborate with urban Amerindians in Manaus in some

---

1   Manaus today, with its nearly 2 million inhabitants, is a city almost entirely built as squatter spaces, spreading outward from the city centre of the port area to the outmost reaches of the city's neighbourhoods over the last 40 years from the free trade era. Like many marginal urban spaces, especially for the urbanized poor of Manaus, there is a connection to an everyday sense of violence or structural violence because of the routinization of poverty and racism (see Farmer 2005: 40; and Scheper-Hughes 1992: 19, 507–8).

way so as to reciprocate the information they would provide me. On this basis, I was introduced to various Indians by a UFAM anthropologist and later by a priest/linguist of the Indigenist Catholic Missionary Council (*Conselho Indigenista Missionário*, CIMI) with the premise of my potentiality as a political ally who might disseminate their cause, their concerns, and their problems to a wider English-speaking public.

At the beginning of fieldwork I was first introduced to a Sateré-Mawé family by my anthropologist colleague – a group, by the way, who had been studied extensively by anthropologists and anthropology students alike at UFAM – and thereby who were accustomed to the ins-and-outs of ethnographic inquiry. My first act of engagement was helping this urban group of Sateré-Mawé repair the roof of their *barracão* (thatched-roof hut) since it had been burned by a racist white neighbour as an alleged warning against their presence in the neighbourhood. (As will be demonstrated below, racism against indigenous peoples in Manaus has bolstered their ethnic identity in defining difference.) My acquaintance with the linguist/priest allowed me to gain access to several other urban Amerindians in the city. These included (aside from Sateré-Mawé): Apurinã, Kambeba, Kokama, Mura, Munduruku, Tikuna, and Tukano.[2] In working with these urban Amerindians I realized soon enough that I was a 'political object' to them, someone with the potential of advocating their views abroad to an international audience and as strategic potential for their cause.[3]

The concept of 'anthropologist as political object' is worth examining in more detail (as I do so below) for it frames my advocacy for urban Amerindian politics in Manaus and their engagement with me as opening political possibilities for them. Since the 1990s the concept of 'engagement' in American anthropological publications has been widely circulated.[4] As a topic of interest, public engagement has a long history of antecedents in cultural anthropology through the works of Franz Boas, Margaret Mead, and Ruth Benedict (Low and Merry 2010) as well

---

2 The priest allowed me to use CIMI offices in Manaus for interviews so some of these groups were contacted there.

3 It is beyond the scope of this chapter to explain the current and past economic, political, and social organization as well as the cosmological myths and world views of these Amerindians. Most secondary reference material about these indigenous peoples represents those Indians living in the interior rather than in the city. In the urban environment the Indians have adapted to changes of living in the city, some keeping customs and language, others losing them.

4 The concept of 'engagement' in anthropology may be attributed to a variety of sources, especially from American anthropologists. Some of these include the following: Wright (1988); Scheper-Hughes (1995); Smith (1999); Kirsch (2002); Lamphere (2003); Hale (2006, 2008); Sanford and Angel-Ajani (2006); Sillitoe (2007); Lassiter (2008); Kellett (2009); Sawyer (2009); Low and Merry (2010); Susser (2010); Johnston (2010); González (2010); Herzfeld (2010); Howell (2010); J.L. Jackson (2010); Spencer (2010); Clarke (2010); Ali (2010); Smart (2010); Mullins (2011); Baer (2012); Linstroth (2012); and Mahmood (2012).

as to some extent in social anthropology through Edmund Leach's efforts to make anthropology more relevant and more pertinent to current affairs (Hannerz 2003). As Sanford and Angel-Ajani (2006: 3) claim: 'anthropologists have raised ethical issues relating to advocacy and politics since the early twentieth century'.

For the purposes of this chapter I limit a definition of anthropological engagement solely in relation to activism and advocacy as it pertains to my own fieldwork and in the manner of my interaction with my informants.[5] My view is borrowed from Charles Hale (2006: 97) wherein I 'affirm a political alignment with an organized group of people in struggle and allow dialogue with them to shape each phase of the process [of research], from conception of the research topic to data collection'. It is a dialogue 'in which anthropologists … offer their services to native struggles' (Wright 1988: 366). Specifically, my understanding is close to Bruce Albert (1997: 56) as he remarks:

> anthropologists find themselves faced with two ethical and political obligations, which were eluded by classical ethnography, but are unquestionable nowadays: on the one hand, being accountable in their work to people who were traditionally only the "objects" of their studies; on the other, assuming the responsibility their knowledge entails for these peoples' resistance strategies vis-à-vis the dominant nation-states' discriminatory and despoiling policies.

Long gone are the days of benign paternalism of white male anthropologists serving colonial administrations over their native subjects, or, for that matter, the 'benevolent autocracy' of the likes of Radcliffe-Brown over the Andamanese (Tomas 1991: 76–7). As Paul Sillitoe (2007: 157) explains: 'it is no longer tenable – if it ever was – for us to represent the lifeways and beliefs of others. Most populations are able to represent themselves'.

When stating that I was considered a 'political object' by the Indians I became acquainted with and worked with, I am asserting the power of their social agency in regard to me and their intentions for my ethnographic fieldwork among them.[6] In assessing this type of engagement I will attempt to theorize what anthropological engagement means in dealing with contemporary indigenous groups who are politically minded and motivated. In doing so I am fully aware that political engagement is fraught with problems whereby it 'is bound to be conflictive and unclear; one has to make decisions, about what to advocate for, on whose behalf, which are basically value judgements, even if these are informed by anthropological knowledge' (Wade 2010: 157). Or, as Jackson and Warren (2005:

---

5 Low and Merry (2010) describe several inclusive definitions for 'engagement'. These are: 1) sharing and support; 2) teaching and public education; 3) social critique; 4) collaboration; 5) advocacy; and 6) activism.

6 Terence Turner (1991: 310) phrases it differently. In referring to the Kayapo Indians of Brazil he says, 'as an anthropologist, in short, I had become a cultural instrument of the people whose culture I was attempting to document'.

556–7) have put it, 'ethnographic practice that bridges inquiry, activism, and participatory approaches to the production of cultural knowledge raises complex questions, epistemological and ethical, answers to which are not exactly around the corner'.

Some anthropologists have even argued that advocacy is basically 'incompatible with anthropology' (Hastrup and Elsass 1990: 301) and lately the Brazilian anthropologist Alcida Rita Ramos (2008: 480–81) has called for a 'disengagement' of anthropologists among natives as 'these "others" are at last affirming their full agency as producers of anthropological knowledge' through autoethnographies (see Nichol, this volume, for more on indigenous peoples' suspicions of anthropological research; also see Sandri, this volume, for an indigenous academic perspective on knowledge production).

In general, theorizing about engagement in anthropology has been widely discussed in North American cultural anthropology but is virtually absent in British social anthropology. Relevant questions today are: what do engagements mean with indigenous peoples (anthropologists and indigenes)? How much political engagement with natives actually compromises anthropological objectivity? Does an anthropologist negate his academic standards in being politically engaged and by taking on a political stance? What does engagement tell us about social agency, intersubjectivities, interethnic dialogues, and reciprocity?

In relegating the power of inquiry to the Brazilian Indians I collaborated with and worked amongst, I allowed them to decide the direction of my research and what my investigation meant to them (see Calabrò, this volume, for indigenous collaboration and participation with anthropology projects). After all anything I published would be given back to them, in the form of articles, book chapters, and eventually in the form of a future monograph. In asserting that these Amazonian Indians living in the city of Manaus have social agency I draw from the conceptualization of Alfred Gell (1998: 16), when he remarks: 'agency is attributable to those persons ... who ... are seen as initiating causal sequences of a particular type, that is, events caused by acts of mind or will or intention, rather than the mere concatenation of physical events'. The social agency of these urban Amerindians exemplifies their considered input in the manner in which an anthropological study could be used to their advantage and therefore their intent in regard to myself. It should be noted such intentions in dealing with anthropologists were nothing new to them. After all, most Brazilian anthropologists they knew provided for them strategic advantages in the past and also in an on-going basis toward government agencies and in the project of their recognition. Hence, in this respect, Brazilian anthropologists in many ways are political objects as well in the politics of indigenism and in indigenist [interethnic] relations.[7]

---

7   Indigenism is actually a complex concept. According to Ramos (1998: 6): 'it is a political phenomenon in the broadest sense of the term. It is not limited to policy making by a state or private concern or to putting indigenist [interethnic] policies into practice ... What the media write and broadcast, novelists create, missionaries reveal, human rights

In recognizing indigenous agency, I am not neglecting my own in this interethnic collaboration (or divide). I was all too aware of the role of anthropologist in the Foucauldian sense of power and knowledge in associating with Indians.[8] Another dilemma was language, what Talal Asad (1986: 156) calls 'unequal languages' and in this case of Portuguese to English and even Amerindian languages to English in regard to both Brazilian Indians and Brazilian anthropologists. For the latter, not many works of Brazilian anthropologists are read by North American or British anthropologists with the exception of anthropologists who actually work in Brazil and those Brazilian anthropologists who have published in English such as Alcida Rita Ramos, Eduardo Viveiros de Castro, and Manuela Carneiro da Cunha and recently Aparecida Vilaça, among others. In addition, and in regard to unequal power relations between the urban Indians and myself, I was also acutely aware of the economic disparities between us (their marginalized and impoverished condition and my relatively well-off status). All of these issues point to ethical dilemmas for American and European anthropologists working with indigenous peoples in the third world and collaborating with non-Anglo anthropologists in regions such as Latin America. Restrepo and Escobar (2005), for example, have suggested a new epistemological approach in thinking of 'world anthropologies' and 'other anthropologies' (knowledge production of other anthropologists around the world) as opposed to the dominant anthropologies: American, British, and French.

Today indigenist relations of anthropologist and Indian are therefore often determined and reinforced by the intentions of the latter but also in the former's competence of aiding Indians, in mutual trust, and sometimes in the bifurcating aims of the relationship. Whereas for anthropologists the purposes of such relationships also encompass academic projects and empirical data-making, while for Indians, at least for the urban Indians in Manaus, there is the politics of being represented, and the representativeness of anthropology acting as political vehicle. As such the special role anthropology and anthropologists play in Indian lives has been a continuously evolving relationship, and in Manaus, Brazil, in concerted commitment for urban Amerindians since at least the 1990s. This is a particular intersubjective association, which as will be explained below, also has

---

activists defend, anthropologists analyse, and Indians deny or corroborate about *the* Indian contributes to an ideological edifice that takes the "Indian issue" as its building block. Lurking behind the images of *the* Indian composed of this kaleidoscopic assortment of viewpoints is always the likeness – or, more appropriately, unlikeness – of *the* Brazilian. Indian as mirror, most often inverted, is … a recurrent metaphor in the interethnic field. In other words, Indigenism is to Brazil what Orientalism is to the West'. For a broader discussion of indigenism and indigenous peoples in terms of a global movement and human rights (see Niezen 2003).

8   As Foucault (1980: 51–2) explains: '… I have been trying to make visible the constant articulation I think there is of power on knowledge and of knowledge on power … The exercise of power perpetually creates knowledge and, conversely knowledge constantly induces effects of power'.

been evident at the national level for determining indigenous affairs. According to Turner (1991: 300–301) contacts of anthropologists with the Kayapo (another Brazilian Indian group) have made the Kayapo aware of 'the potential political value of their "culture" in their relations with the alien society ...'. Anthropologists have in essence taught Indians, and in this case Brazilian Indians, to value their culture, which has had a tremendous effect on the history of indigenous politics in Brazil from the beginning (at the national level since the 1970s).[9] Although local politics, as anywhere else, are determined by local issues and those in Manaus concerning urban Indians are particular to the problems occurring in the city, the national indigenous movement at times influences the local, especially in forming Pan-Amerindian identities. To be situated in the city of Manaus in the state of Amazonas as an anthropologist with Brazilian urban Amerindians is to be situated in their politics of recognition, which means taking a political stance against the discrimination and racism of white society (*brancos*) and against governmental agencies such as the National Indian Foundation (Fundação Nacional do Índio, FUNAI) and the National Health Foundation (Fundação Nacional de Saúde, FUNASA). These local Indian politics mirror national indigenous politics in Brazil as indigenous struggles for recognition exist with Brazilian society in general and against government agencies all over Brazil.

My engagement with these urban Indians meant in many ways that I would act as their surrogate in expressing their concerns to foreign English-speaking audiences, a role that anthropologists know all too well. Yet in doing so, I was supposed to raise awareness about recognizing their indigenous rights, which may in turn pressure the Brazilian government to officially acknowledge their presence, act on their behalf, and to have the same rights as Indians in the interior of Amazonia. Hence, this chapter will be about expressing urban Amerindians' concerns about the discrimination and racism they face against Brazilian society and for many reasons against the Brazilian government. Anthropological engagement of this sort is about exposing past wrongs of indigenous peoples and their ongoing conflicts with the non-indigenous population for more than 500 years.

Yet anthropological engagement of this type also entails something else, it is a give-and-take arrangement. It is about reciprocity.[10] This is an exchange relationship between Indian and anthropologist over broader political recognition for the former

---

9   Even so, it was really when the Catholic Church began organizing indigenous meetings throughout the country beginning in the 1970s that the indigenous movement in Brazil began in earnest with of course the support of Brazilian anthropologists.

10   In discussing reciprocity and collaboration with indigenous peoples, Hale (2008: 503) maintains that: 'three specific conditions of possibility constituted the emergence of reciprocal relations of collaboration between white anthropologists and indigenous peoples: rising indigenous militancy in national level struggles for collective rights; racial tensions between Indian-and-mestizo-led political initiatives; and changes in the sensibilities of US based anthropology toward a special emphasis on close, horizontal relations with 'subaltern' research subjects'.

and information for the latter. These are values, which emerge in 'action', such as indigenous peoples providing the anthropologist with information whereby he/she will in turn act through publication or through other forms of mediation (film, internet web sites, etc.). The value is through the 'act of giving' one to the other in an 'open' form of reciprocity which 'implies a relation of permanent mutual commitment' (Graeber 2001: 45, 220). The desire of the anthropologist as political object is a strategic one for these urban Indians for they know that anthropological knowledge is a powerful tool and the anthropologist himself/herself may provide access to NGOs, government agencies, and the media to their cause. Such a relationship is not materialistic in contrast to Amerindians coveting Western goods as is common throughout Amazonia (see Hugh-Jones 1992). It is, however, about sharing and somewhat about 'good gifting' common among Amerindians throughout Lowland South America (Overing and Passes 2000: xiii–xiv). It is an interethnic relationship in the manner of an informal contract. Field research of this kind of anthropological engagement makes known the 'unrepresented' and underrepresented (Dresch and James 2000: 23). Engagement of the anthropological variety is enmeshed with the mutually reaffirming intersubjectivities of a special type of interethnic relationship, while compromising objectivity it takes on moral concerns for fairly representing indigenous agency and their intentions for anthropological outcomes.[11]

The purpose of this chapter therefore is to provide an overview of urban Amerindian identity in Manaus, Brazil, and their struggles for recognition (see Sillitoe, this volume, for more on 'representativeness of indigenous representatives').[12] In this discussion I will analyse the autobiographical memories of discrimination and racism of urban Amerindians and explain how these traumatic memories are particular for indigenous peoples in righting historical wrongs and in signifying their indigeneity in contrast to Brazilian society as well as point to some theoretical interests about indigenous trauma. My analyses of memory move beyond what Charles Hale (2006: 113) has termed the 'social memory of struggle'. In my view indigenous peoples and indigenous movements base part of their struggle for recognition on memories of racism and discrimination. These are not simply political memories passed on to successive generations. Such memories as I will describe below exhibit the particular interethnic and racial encounter indigenous people have suffered for more than 500 years. In sum, memories of trauma serve to delimit them from white society (*os brancos*) and express their

---

11   For more on 'intersubjectivities' and 'subjectivity' see Biehl, Good, and Kleinman (2007).

12   Part of the political struggle of urban Amerindians in Manaus such as recognition of greater health rights is left out of this discussion for the purpose of length. Additionally, this discussion does not link other indigenous movements throughout Latin America to Manauran indigenous politics or even Brazilian Pan-Amerindian politics because of word limitations. Such discussions will be published elsewhere.

identity differences from the non-indigenous. Indeed this microcosm of memories of discrimination and racism are global indigenous concerns as well.

In the next section, I will describe the urban Amerindian population living in Manaus, Brazil and their socio-cultural characteristics.

Figure 6.1   Tuxaua (leader) Paulo, Sateré-Mawé, dressed up in full regalia in the city of Manaus 2009

Note: Such urban indigenous dress on special occasions is not traditional but used as a strategy for recognition.

Urban Amerindian Population in Manaus

At present, there are approximately 896,917 indigenous peoples living in Brazil (0.47 per cent of the total population) (IBGE, Instituto Brasileiro de Geografia e Estatística 2010). Of this number approximately 572,083 of them still live traditionally on 612 indigenous reserves (from 291 ethnic groups, speaking 180 different languages, on approximately 106 million hectares of land or 12 per cent of the total Brazilian land mass). The remainder of these indigenous peoples do not belong to any particular indigenous reserve in the hinterland and some reside in urban areas (approximately 324,834) (see Hemming 2003; Luciano 2006; IBGE 2010).

In Manaus there are between 15,000 and 35,000 Amerindians (Berno de Almeida 2008b: 32).[13] Amerindians move from the interior to the city for some of the following reasons: for better economic opportunities, including comforts such as electricity and running water; for better education for their children; for health reasons and access to medical facilities; for social relations to be with affinal and consanguineal kin; and lastly, because of unhappiness on the reservation and social problems. Urban Amerindians are not completely cut off from their kin in the Amazonian interior. Most make trips back to their village home communities to visit relatives at least once a year. Some visit every three or four months and relatives of the Amazonian interior are frequent visitors to their urban Amerindian relatives.[14]

One may ask how do Amerindians survive in the city? Of the urban Amerindians I met, they are resourceful in pooling money. Indigenous communities in the city also help new arrivals from the interior by providing lodging and food for new families. Common strategies for surviving in the city include selling artisanal crafts and some of the husbands of indigenous women work as day labourers, often on construction jobs.[15] The other principal source of income is the Brazilian 'Family Stipend' (*Bolsa Familia*), which is a social welfare fund to help impoverished families.

We may think of urban Amerindians in terms of their 'emergent identities' because of the variation among them as newcomers or long-term residents of the city (Oliveira Filho 1999a: 107). Their identities shift and change with their community associations and marriages and one must not think of their ethnicity as having any primordial or fixed qualities whatsoever (see Banks 1996; Jenkins 1997). In Manaus one may encounter second and third generations of Amerindians who only speak Portuguese, and still others who are fluent in both Portuguese and their native tongues. There are no strict marriage rules per se for exogamy among the urban Amerindians as in the past. Whereas for exogamous unions one may find marriages between Munduruku and Sateré-Mawé of Southern Amazonia

---

13   An IBGE 2010 estimate for Indians in Manaus was only 3,837.

14   According to FUNAI in 2008 the following ethnic indigenous groups (altogether 62 groups) live in Manaus (although some groups are more politically visible than others): Apurinã, Issé, Katawixi, Marimam, Parintintin, Tuyúca, Arapáso, Jarawara, Katukina, Marubo, Paumari, Waimiri-Atroari, Juma, Katwena, Matis, Pirahã, Wai-wái, Banavá-Jafí, Juriti-Tapuia, Kaxarari, Mawaiâna, Pira-Tapúya, Wanana, Baniwa, Kaixana, Kaxinawá, Sateré-Mawé, Warekena, Barasána, Kambeba, Kaxhysana, Siriána, Wayampi, Baré, Kanamari, Kobema, Mayoruna, Tariána, Xeréu, Deni, Kanamanti, Kokama, Miranha, Tenharin, Jamamadi, Dessana, Karafawyána, Korubo, Miriti-Tapuia, Torá, Yanomami, Hi-Marimã, Karapanã, Kulina/Madijá, Munduruku, Tukano, Zuruahã, Hixkaryana, Karipuna, Maku, Mura, and Tikuna (Maximiano 2008: 97).

15   Many husbands of indigenous women are *caboclo* or *pardo* (of indigenous descent but not necessarily self-identified as such) and are not as active in the indigenous affairs of the community. (These men are mostly identified as whites – *brancos* instead of *pardos* and *caboclos* – either because of their unawareness of the exactness of their indigenous origins, or an unwillingness to be associated with being Indian.)

(rather than between the Sateré-Mawé's own patrilineal clans for example) and in contrast to the exogamous unions between the different groups of Indians from Northwest Amazonia.[16] There are many marriage unions between *caboclos* (whites of indigenous descent) and Indians and many children out of wedlock as well and even between different partners.[17] Some of the urban Amerindians practice Seventh Day Adventism, others Catholicism, and still others Baptism, or some other Protestant Pentecostal religion (see Vilaça and Wright 2009 for an elaboration about Amerindians and Christianity in Brazil). Most of these self-identified Amerindians living in the city are not recognized by FUNAI simply because Indians are only considered to be Indian if they live in the Amazonian interior. Once an Amerindian person moves or lives in the city they are not officially recognized because stereotypically they have become 'civilized' (*civilizados*) according to FUNAI and Brazilian society.[18] This also means that they do not share all the same benefits as their indigenous kin of the interior.

Figure 6.2   Brazilian military police accosting Iliana, Sateré-Mawé, at Lago Azul occupation site, 11 March 2008

16   Marriage rules among the different groups are known but mostly in the past and in the interior.

17   A *caboclo* is an ethnic term for the admixture of Indians with whites.

18   Warren has also discovered that FUNAI does not assist Indians in the Southeast of Brazil for the same reasons. As he states (2001: 113): '… I am arguing that the government assistance is not substantive. As we have learned, even in those cases where significant amounts of money have been allocated for certain services, once funneled through FUNAI the "real" benefits are scant. Moreover, the social services that actually materialize tend to do so in such a haphazard, distorted, and ineffective manner that it is often a stretch to refer to them as benefits'.

The urban Indians of Manaus are primarily organized by their various communities and associations representing various indigenous groups such as Tukanos or Sateré-Mawé.[19] Such associations and communities are not devoid of interethnic problems and political infighting between them. There is continual disagreement among such indigenous communities but when confronting the Brazilian government all ethnic groups come together and protest as a whole in their confrontations as a Pan-Amerindian movement.

The politics of recognition for urban Indians is closely entwined with their memories of racism and discrimination they have experienced in Manaus. As will be described below, such memories also become an impetus for indigenous action and the formation of political movements.

Memories of Discrimination and Racism among Urban Amerindians

To address issues of discrimination and racism in the urban environment of Manaus, it is important to discuss the demographic significance of the population. In Amazonas state about 70 per cent of the population is considered to be *parda* or euphemistically 'brown', a very broad racial term without considering ancestry or descent (IBGE 2010).[20] It is safe to state that the overwhelming number of *pardos*

---

19 Some of these indigenous associations in Manaus are the following: AACIAM – *Associação de Arte e Cultura Indígena do Amazonas*; ACIBRIN – *Associação das Comunidades Indígenas do Rio Negro*; ACIMRN – *Associação das Comunidades Indígenas do Médio Rio Negro*; ACINCTP -- *Associação Comunitária Indígena Agrícola Nhengatu*; ACIRI – *Associação das Comunidades Indígenas do Rio Içana*; ACIRU – *Associação das Comunidades Indígenas do Rio Umari*; ACIRX – *Associação das Comunidades Indígenas do Rio Xié*; ACITRUT – *Associação das Comunidades Indígenas de Taracuá, Rio Uapés e Tiquié*; AIM – *Associação dos Indígenas Munduruku*; AISMA – *Associação Indígena Sateré-Mawé do Rio Andirá*; AMAI – *Associação das Mulheres de Assunção do Rio Içana*; AMARN – *Associação das Mulheres Indígenas do Rio Negro*; AMIK – *Associação das Mulheres Indígenas Kambeba*; AMISM – *Associação das Mulheres Indígenas Sateré-Mawé*; AMITRUT – *Associação das Mulheres Indígenas de Taracuá, Rio Uapés e Tiquié*; APN – *Associação Poterihnarã-numiá*; CAFI – *Centro Amazônico de Formação Indígena*; CGTSM – *Conselho Geral da Tribo Sateré-Mawé*; CGTT – *Conselho Gerald da Tribo Ticuna*; CGTT – *Conselho Geral da Tribo Ticuna*; CIKOM -- Coordenação *Indígena Kokama de Manaus*; CIM – *Conselho Indígena Mura*; CIMAT – *Conselho Indígena Munduruku do Alto Tapajós*; MIK – *Movimento Indígena Kambeba*; MPIVJ – *Movimento dos Povos Indígenas do Vale do Juruá*; OIBI – *Organização Indígena da Bacia do Rio Içana*; OPAMP – *Organização do Povo Apurinã da Bacia do Rio Purus*; OPIAM – *Organização dos Povos Indígenas do Alto Madeira*; OPIMP – *Organização dos Povos Indígenas do Médio Purus*; OPIPAM – *Organização dos Povos Indígenas Parintintin do Amazonas*; OPITTAMP – *Organização dos Povos Indígenas Torá, Tenharim, Apurinã, Mura, Parintintin* e *Pirahã* (Berno de Almeida and Sales dos Santos 2008: 9–10).

20 In the state of Amazonas 68.9 per cent are *parda* (IBGE 2010). Few authors have discussed racism in relation to Brazilian Amerindians but most notably are Oliveira Filho

in the state correspond with a *parda* population in Manaus where the majority in the state reside.[21] It may likely be assumed that many of these *pardos* are actually also *caboclos* because of the prevalence of indigenous descent in the region and because of successive migrations to the city from the Amazonian interior of Brazil since the 1970s from the 'Free Port' era.[22]

According to Jonathan Warren (2001: 339) a *caboclo* is 'an individual of predominant or salient indigenous descent who in cultural terms is considered to be only loosely connected to Indians (a "detribalized" Indian)'. Yet such persons present certain paradoxes as much of the population in Manaus embrace white culture whilst negating their indigenous heritage.[23] As Alcida Rita Ramos (1998: 77) asserts: 'The *caboclo* is the embodiment of the paradox contained in the civilizing project: the effort to wipe out Indianness while closing the doors to their full citizenship. What is left in the wake of such ambivalence seems to be no one's concern'. Notions of racism are passed on from the colonial era of the

---

(1998, 1999a, 1999b, 1999c, 2004); Ramos (1998); Albert and Ramos (2000); and Warren (2001). Most authors who write about racism in Brazil discuss Afro-Brazilians. Some of these authors include: Bailey (2009); Davis (1999); Hanchard (1994, 1999); Marx (1998); Nascimento (2007); Ribeiro (2000); Sansone (2003); Seigel (2009); Sheriff (2001); and Twine (1998).

21   As Oliveira Filho (1999b: 128–9) rightly points out the census category for *pardo* is 'problematic', because it is 'an operational category – artificial, arbitrary, and with appearance of being technically scientific'. He goes on to state: 'The census category of "pardo" pretends to indicate exactly – for the possibility of its measurement – a situation of mixture between different groups of colour. If it is a primordial objective, it is nonetheless pointing to the existence of mixture – or that is an intersection between different categories – it is therefore then only possible to understand the reasoning for not including the various types of mestizos' (Oliveira Filho 1999b: 135).

22   After the rubber boom era in the nineteenth century, Manaus did not experience an economic resurgence again until 1967 with the Brazilian government proclaiming it a 'free port' (*Zona Franca*), meaning Brazilians could fly to the city to buy tax-free imported goods and 'manufacturers enjoyed generous tax concessions' (Hemming 2003: 460). People moved to Manaus in droves because of this new economic stimulus. 'As a result Manaus grew explosively from some 150,000 people in 1965, to 600,000 a decade later, and 1,100,000 at the end of the [twentieth] century' (Hemming 2003: 460). As Ferreira de Melo and Freitas Pinto (2003: 46) explain, new migrants to the city have unrealistic expectations about Manaus as imagining a place with 'many options' and have numerous fantasies about the special merits of the urban environment prior to moving to the city.

23   Local society magazines in Manaus such as *Top Line* actually promote white identity and photographically whiten the complexions of some of the society photos to make people have lighter skin colour. Such news media focus on whiteness and white beauty is dominant in Brazilian society. For an Amazonian city like Manaus the promotion of white aesthetics, including the consumption of material goods, makes it similar to other Brazilian cities. What is dissimilar is Manaus' history of the rubber boom era and the negative attitudes in general toward Indians. The irony of course is that a good proportion of the population in Manaus have indigenous ancestry.

'rubber boom' to the present, making the negation of one's indigenous ancestry in Manaus something of the norm. Amerindians to most Manauarans are considered in stereotypical terminology as 'backward', 'lazy', 'uncivilized', 'children', 'heathens', 'nomads', 'primitives', and as 'savages' (see Oliveira Filho 1998: 61; Ramos 1998: 15–47).[24] Such discriminatory classification was evidenced by the many conversations my wife and I had with the so-called non-indigenous in Manaus. Rejecting indigeneity was normal so as not to suffer the stigmas and barbs of racism of everyday city life for being associated with Indianness or being an Indian. If 'the Brazilians are civilized – the Indians should be wild. Anything in between is sheer pretense. These *caboclos* are Indians who play at being *brancos* [whites] but convince nobody' (Ramos 1998: 77). Manaus is an indigenous city and a space where *pardos* and *caboclos* are the majority of the population. Manaus is also an urban landscape of negation, of denying indigenous heritage, where self-declared Indians must assert and reinforce their identities in order to be heard and recognized by the majority.

It is in this atmosphere where one encounters the urban Amerindian of Manaus. During my time in Manaus, and as I mentioned above, I was able to interview Amerindians from eight different ethnic groups: Apurinã, Kambeba, Kokama, Munduruku, Mura, Sateré-Mawé, Tikuna, and Tukano. All of the people representing these indigenous groups told me their stories about racism and discrimination they experienced in the city. Aside from the usual stereotypes of Amerindians in Brazilian society, to many Brazilians the very idea and image of the urban Indian is similar in many ways to the discriminatory view about urban Indians in the United States:

> The words "urban Indian" immediately bring to mind an image of a poor full-blood Indian living in a city who has been victimized by urbanization. He is undereducated or lacks sufficient training or skills, and seems out of place in the city. Instantaneously the "Indian" is identified with deprived African-Americans and underprivileged Hispanic-Americans. His home is envisioned as one similar to the projects in ghettoes or dilapidated apartments in barrios. (Fixico 2000: 26–7)

In Brazil such a stereotypical vision may be replaced with Afro-Brazilian and ghettoes with *favelas*. As Warren (2001: 173) declares:

> Indians, then, are not imagined as catching the subway, drinking soda, piloting airplanes, using credit cards, watching television, and so on. They are also *not* thought of as being doctors, college students, janitors, maids, factory workers, or lawyers. Indians are not considered to be residents of urban shantytowns, beachfront resorts, suburban homes, or plantation estates. To live in these so-

---

24   For urban Amerindians, they were considered to be 'civilized', 'detribalized', and 'mixed', adding to the pejoratives about them see also Oliveira Filho (2004).

called civilized spaces, to be in these allegedly modern occupations, to possess the latest consumer goods of the global economy, renders someone non-Indian. Such a person is racially positioned as moreno, pardo, or Asian – but not Indian. To be an authentic Indian, one must live like a primitive in a traditional manner. One must embody the antiself of civilization, which in Brazil means living in a hut in the middle of the forest, naked, and with no contemporary technological conveniences.

In Manaus, urban Amerindians, as Berno de Almeida (2008a: 14–15) declares, are part of the poor and excluded populations of the city but we must be careful not to treat them as the 'exoticized poor' (see also Bernal 2009). Warren (2001: 21) describes the Indians he encountered in southeast Brazil as 'posttraditional' as 'the experience of the dramatic shattering of tradition and to refer to a longing, an orientation, that involves an active attempt to rediscover, recuperate, and reinvigorate that which has been dismembered'. Yet many urban Amerindians in Manaus are not simply Amerindians trying to revive their culture but also ones who wish to maintain their culture within an urban environment hostile to the notion of indigeneity. In Manaus some second and third generation Amerindians want to learn their native language and are active participants in the CIMI programs of language revitalization, while other Indians are fluent in their native tongues and try to create spaces, such as women's associations, in which to talk in their language and make artisanal crafts. The pressures in Manaus not only follow the national stereotypes admiring whiteness and white culture but the discrimination experienced by most self-declared urban Amerindians is one harking back to the rubber boom past. The colonial past and the rubber boom era in the late nineteenth century and early twentieth century created a 'space of death', 'where torture is endemic and where the culture of terror flourishes' (Taussig 1987: 4). The debt-peonage system enslaved Indians as part of forced labour, creating an 'economy of terror' and had a deleterious effect whereby 'rites of conquest and colony formation, mystiques of race and power ... instead bound Indian understandings of white understandings of Indians to white understandings of Indian understandings of whites' (Taussig 1987: 51, 109). In other words, white fears of so-called Indian savagery were transformed by whites into white savagery through the civilizing abuses and terror by whites against Indians as part of the rubber economy. The white imagination about Indians therefore idealized a commitment by whites to grotesque acts of violence against Indians thereby perverting and reversing the role of savagery as the transformation of the rubber traders and rubber barons into monsters. It may be argued such past visions of racism about Amerindians are still in the present and form part of the stereotypical vision of many non-Indians in Manaus. This is why I argue that the discrimination and racism experienced by urban Amerindians today is a form of 'negative circulation' at once embodying the rubber boom past in the present and at the same time as reifying the negative stereotypes of anti-civilization and the opposition of Indianness to white society. Partially too, the vision embodies nationalist ideals of positivism and the origins of Brazilian nationalism and the civilizing goals of the Brazilian nation as suggested by the motto on the national flag 'Order and Progress' (*Ordem e Progresso*) (Hemming 2003: 14–15).

When confronted with 'white' society, attending meetings, conferences, or other public venues, such as protests, most urban Amerindians I knew would dress up and adorn themselves by painting themselves and the leaders would wear their 'headdresses' (*cocares*) and other feathered and artisanal adornments (see Graham 2002; and Conklin 1997).[25] At official meetings and in public venues this was especially important, as the Amerindians were able to demarcate and embody their indigeneity in an urban environment hostile to Indianness. This was a strategy to negate the circulation of discrimination but also to celebrate their indigeneity.[26] On 19 April, or the 'Day of the Indian' (*Dia do Índio*) (a national commemoration), and on other festive occasions, were periods in time when urban Amerindians felt obliged to dress up and differentiate themselves by displaying their self-declared status as Indians in Manaus. As Oliveira Filho (2004: 28) articulates this is a 'political ritual where it is always necessary to demarcate the borders between "Indians" and "whites"'.[27] Hence, displaying one's indigeneity with paint and adornments may be considered as part of the 'neo-indigeneities', as the new forms of embodying indigenous ideals when confronted with the white world. Indeed, making the distinction between the Indian world and the white world was ever present in the narratives of the urban Amerindians I spoke to on a regular basis. Many of the Sateré-Mawé I knew, for example, referred to 'whites' as *cariuá* (or *kariwá*).[28] They described whites as: 'to themselves', 'each to their own', 'not sharing', 'uncaring', 'selfish' – and complained about the crime in the city of Manaus. To many indigenous people life in the city was much more difficult than the Amazonian interior because of the need for money and the way people acted toward them.

In my view I would describe and argue that urban Amerindians living in Manaus are 'post-modern Indians' in contrast to 'posttraditional Indians', those Indians with an admixture of non-indigenous heritage who have also lost traditional customs and language (Warren 2001), or 'hyperreal Indians', the idealization of an Indian to be in a pristine and noble state perpetuated by some NGOs and others (Ramos 1994, 1998). The reason urban Indians living in Manaus are 'post-modern'

---

25   It should be reiterated that so-called white society in Manaus is really the majority *parda* population, many of whom are really *caboclos*.

26   In personal communication with Wolfgang Kapfhammer, he likewise affirms this strategy of urban Amerindian regalia.

27   Many times during my interviews, the urban Amerindians would feel it necessary to dress up, especially if I was filming them. For more perspectives on dress and Amazonian indigeneity, see Ewart and O'Hanlan (2007).

28   In Tupi-Guarani white men or the white race is simply *cari*. These words are probably derived from the *lingua geral* (general Tupi language), *Ñe'engatú*. This language was standardized by the Jesuits in the sixteenth and seventeenth centuries in the Brazilian Amazon.

is that they accept certain aspects of modern life (e.g. televisions, VCRs, taking the bus, using the internet) but reject other aspects of it, the selfishness of whites for example. They may ascribe to wearing headdresses, body ornament, and body paint at meetings and conferences with whites or at protests against whites but then return to Western dress in their everyday lives.[29] To outsiders (non-Indians) urban Indians were often regarded as impostors or as civilized, as mentioned, because of their acceptance of modernity. Indians living in the city, as I knew them, critiqued white society and were able to deconstruct it in their terms as anti-Indian and many times as racist. Their acceptance of some aspects but not others of Brazilian society made them able to navigate their identity in innovative ways, a deconstructive position, and to outsiders as the anomaly of Indian in the city. Urban Amerindians often played into their anomalous status through their associations and communities. As many of them kept their customs such as the *tucandeira* ritual for the urban Sateré-Mawé and in many cases their native languages did not make them posttraditional but rather they are repeatedly reifying alterities of novel indigeneities through self-aware contrasts and strategies.[30] Urban Amerindians may be traditional yet modern and at the same time deconstruct white society as inferior to Indianness and even manipulate some aspects of white society to their political gains as in 'anthropologists as political objects'.

The following are excerpts from interviews I conducted with several urban Amerindians in Manaus in 2009.[31] Such views encapsulate the varying experiences of racism by urban Amerindians and exemplify different Indian sentiments about being discriminated against during different times in their lives. It may be argued that some of the trauma of discrimination may be regarded as episodic and as poignant reminders of particular difficult periods in the past and other trauma forms of discrimination as having an everyday quality, as being continual, chronic, and part of living in the city. I referred to these types of trauma elsewhere as 'synchronic trauma' (episodic) and 'diachronic trauma' (recurrent) because of the cognitive effects of memory on the individual and how traumatic events access different memory forms via time, particularity, and with acute and chronic qualities (Linstroth 2009: 168–70). Both synchronic and diachronic traumas in reality exemplify one type of trauma as a whole. While racism against an Indian individual is described as occurring at a particular point in time (synchronic), all forms of racism against Indians have developed and continue over time (diachronically). Memories of discrimination and racism therefore are remembered as much for their particularity at a given moment in the past as for their re-occurrence over time.

---

29   Other anthropologists have also commented on how Indians manipulate their images for white society and their uses of modern Western technologies as relational strategies (see Conklin 1997; Turner 2002).

30   The *tucandeira* dance is a rite-of-passage ritual for young men to dance with gloved hands containing hundreds of stinging bullet ants (*Paraponera clavata*).

31   All the names of interviewees are pseudonyms in order to protect the identity of Amerindian informants.

In quoting below urban Amerindian memories of discrimination and racism I am referring to them as autobiographical memories following Maurice Bloch (1998: 116) and in particular how 'people recall the past, either in the presence of others or in the imagined presence of others, in heard or silent soliloquies'. However, recently following Bloch's (1998) book on cognition and memory, some psychologists have since clarified what is meant by autobiographical memory in a more precise manner than Bloch. According to Conway (2002: 53):

> Autobiographical knowledge represents the experienced self (or the "me"), is always accessed by its content and, when accessed, does not necessarily give rise to recollective experience. Instead, recollective experience occurs when autobiographical knowledge retains access to associated episodic memories.

The narratives below are all exemplars of autobiographical knowledge through episodic memories in relation to the racism urban Amerindians have experienced while living amidst white Brazilian society.

*   *   *

Interview with Paulo (See Figure 6.1), 29 years old, *Tuxaua*, a leader of one group of Sateré-Mawé in Manaus, 23 March 2009

> Paulo: Other Indians when they arrived in that epoch [1960s–1970s] hid their Indian identity, but we never negated our race as Indians, we always remained Indians, never negating our race ... we kept learning and this struggle to be indigenous, we learned about the law ... behind this struggle and our struggle is always *Tupana* [God].
>
> JPL: How do you view the "white" world (*os brancos*)?
>
> Paulo: In this world to each their own, no help ... there is no help, individualism, the indigenous world is not like this. It is organized another way to help one's relatives.
>
> JPL: How do you view discrimination in Manaus?
>
> Paulo: You feel sad, it hurts the spirit, it makes you sad ... but given time and once you better understand it, it makes you want to fight harder ... you use it as a medium to fight. They say to us: "what are you doing here, living in the slums, you are Indians" ... but we have our artisanal crafts, and our culture, the *tucandeira* dance ... so we began inviting the public [to the *tucandeira*], especially public schools. There were few people at first but this increased to more. Those people who discriminated against us began to think differently. Now they begin to value the Sateré-Mawé culture. We do this to demonstrate the Sateré-Mawé culture, for people to understand it. It has become much better but it [discrimination] still exists.

It is clear Paulo's narrative describes how he and other Sateré-Mawé of his family maintained their identity despite discrimination. He reminds us how he and his family migrated to the city but kept their Amerindian identity. His narrative also describes how he as a leader found a way to make the Sateré-Mawé more acceptable to the broader public by holding a sacred rite-of-passage ritual, the *tucandeira* dance, and opening it to the public. By allowing others to see the sacred ritual then people understand their culture. Despite this in March 2009 (as mentioned in the introduction) some unknown neighbours burned the roof of the *barracão* (thatched-roof hut structure), which is a poignant reminder of the discrimination felt by his community, *Comunidade Andirá* and their struggles with city life.

Interview with Davi, 27 years old, Sateré-Mawé, 24 March 2009

They say the Indians belong in the forest, this is what they tell us … [commenting on discriminatory remarks he has heard].

Interview with Yolanda, 20 years old, Sateré-Mawé, 24 March 2009

There was a time when we were very discriminated against … Yes when we were in school, they [other students] used to call us names and the neighbours too but not now, no … the white man does not think about people …

Interview with Iliana, 23 years old, Sateré-Mawé, 25 March 2009

When I was little there was discrimination at school. We heard things like "there are the Indians, they belong in the forest", and with the teacher too … One time I was told by a teacher: "Indians are not worth anything", people saying things like that to me. Those are the things I remember.

Interview with Raquel, 27 years old, Munduruku, married to a Sateré-Mawé, 30 March 2009

Our children are very discriminated against when they go to school, especially when they go to school painted [adornment during their festivities] and this causes shame … then after a while we started having our festival [*tucandeira* ritual] here … then they discriminated against us less.

Davi, Yolanda, Iliana, and Raquel's narratives tell us different stories. Davi comments upon painful remarks of discrimination about Amerindians belonging in the forest. Such verbal taunts reflect the painful memories many urban-Amerindians have to contend with while living in the city. Yolanda also remembers when the neighbours called her names and when she was teased at school as a child. Iliana also relates to a painful memory she experienced in school. Raquel discusses how sometimes her children are discriminated against at school. Such memories have both synchronic and diachronic qualities. That is, patterns of discrimination and their associated memories are particular to certain periods of time – episodes – in opposition to some memories, which are formed from an everyday, continual process of discriminatory actions. Yet whether memories are episodic or part of a continual occurrence of living in the city as Indian, are really exemplars of the same, memories of discrimination and racism. These are memories experienced from particular times and are acutely remembered but the discrimination such as taunting by neighbours and at school must have been usual occurrences that each of them (or their children) had to endure on a regular basis. This is why I have argued that memories may have synchronic and diachronic effects of trauma (Linstroth 2009).

Interview with Iliana (See Figure 6.2), 15 April 2009

[Describing the incident which occurred at Lago Azul on 11 March 2008; the Indians occupied a piece of land near Manaus as homesteaders. They were called 'invaders' and were forced off the land by the Brazilian Military Police.]

> At that moment I was angry ... I was pregnant with child and holding my baby when they [military police] started pushing me ... I was angry because they [military police] were choking my husband with a baton ... they hit me on the arm and started pushing me ... they [military police] said to me: "You are not an Indian, what are you doing here?" "'You are not an Indian". They [military police] said the people [Indians] were "civilized", and that people [Indians] were invaders ... They [military police] came to throw us off the land. I was angry when they hit my arm ... they told us to leave ... that photo appeared in the newspaper [referring to a photo of the incident] ... this occurred on 11th of March, 2008 ... they need to respect the Indians and what they say ... I remember crying ... we had stayed there 10 days.

It was reported that about 150 military police (*Polícia Militar*, PM) armed with batons, shields, rifles, tear gas, and pepper spray dispersed the 200 Amerindians staking claim to the vacant land known as Lago Azul II on the day of 11 March 2008. Many of the Indians were arrested that day as well, including some non-indigenous, trying to stake claims to the vacant land area of approximately 180,000 m². The newspapers reported it as an 'Indian invasion' of vacant lands with some 17 ethnic groups participating from Manaus including: Tikuna, Dessana, Sateré-Mawé, Kanamari, Kokama, and Baniwa (see Dias da Costa 2008: 213–24). The actions by the police against the Amerindian land grabbers reached national attention with *TV Globo* and many newspapers in Manaus reporting the event. Iliana's photo appeared prominently in the newspaper as she fought with the police while carrying her small child and while pregnant with another child. Such police abuse is symptomatic of the general feelings against Amerindians living in the city in general as 'detribalized', 'civilized', 'landless' and so forth as described above. Eventually, FUNAI investigated the incident concerning whether or not police abuse had occurred. As a result the Amazonas State government paid 59 families compensation for the trouble caused by the police with a value of R$10,000 (approx. $4,600 USD today) (Dias da Costa 2008: 224). Clearly though Iliana's story encapsulates the trauma and suffering of enduring police brutality on 11 March 2008.

Interview with Alberto, 52 years old, Apurinã, 4 May 2009

[He describes how he was hit when he was young by non-indigenous people in Manaus trying to survive in the city. His father had died and his uncle left him in the city during his adolescence to fend for himself and find his way.]

> I, I suffered … In Apurinã [culture] when they do not like you they leave you … They [his uncle and other kin] took me to the white city [Manaus], all to each his own … In the streets they called me dog and hit me, paff! … Listen, I suffered a great amount … I learned how to work. I did not know where to go … what happened to my father and grandfather, they suffered too [also, relates to the oppressive work during the Rubber Boom era and the work of his grandfather] … Now it is better and today people in the street treat me with respect … I have a presence in front of white people … many people think of the Apurinã as in the past [referring to past discrimination] … To remember is suffering, a lot of suffering, you do not know which road to take … during my adolescent years I went through a lot of suffering … For me things have gotten much better with work, family, learning …

Of the people I interviewed, Alberto experienced some of the worst suffering imaginable. As an adolescent he was orphaned in the city and had to fend for himself as a street child. He related to me stories how he used to be beaten and how much he suffered. His memories also extend to his grandparent's generation and the rubber boom era when his people, the Apurinã, suffered trauma from that time.

Interview with Neva, 26 years old, identifies as being Kokama with her husband's group, her mother is Apurinã and father is *caboclo*, 13 May 2009

> Unfortunately, there still exists prejudice today … In the past we were like Blacks, now no. It is very awful prejudice and very ugly also … we are also

prejudiced by the government [FUNASA] because we do not get proper health care from them ... Marriages between Indians and Indians is good because the whites do not understand our culture unfortunately.

Interview with Estefania, 22 years, Kokama, her father is Mura, mother is Kokama, 13 May 2009

I see discrimination as really awful because we are Indians. Discrimination is ugly ... discrimination is a big thing. These things are opening how people are thinking. There are more opportunities today ... the whites discriminate against the Indians still ...

The story of these two Kokama women describes their views on discrimination and as still existing for the urban Amerindians. Like many of the other informants, most describe prejudice being better in today's times than in the past. Even so, they recognize differences between what they call the white world and Indian culture.

Interview with Lara, 19 years, Tukano, 29 May 2009

When I was a child there was a lot of discrimination at the school. Classmates did not look at me in a nice manner while in school but things have begun to change as I have been involved in the indigenous movement ... [asked her to explain a personal incident]. When I was in the primary, public state school, there was a teacher who knew I was indigenous. She was a traditional teacher, a white teacher. She told me I did not understand anything because I did not speak Portuguese very well. So she did not treat me very well and neither did my classmates. Then there were fights with my classmates and then after that I did not identify with being indigenous. When I learned Portuguese it was better ... it happened when I was 9 years old ...

Lara's experience with discrimination is another childhood memory, which happened a decade earlier. She was not treated well by the teacher because she did not speak Portuguese well. She also negated her indigenous identity for a while out of shame. Such incidents, I believe, also enforce notions urban Amerindians have about white society and the majority population. Such suffering also emboldens indigenous identity as well.

*   *   *

In summarizing these memories of racism and discrimination as experienced by urban Indians in Manaus, I associate their recollections with a certain type of trauma known to indigenous peoples throughout the world for more than 500 years since colonial times and continuing today in postcolonial states. However, the word trauma is an evocative concept and really a recent invention by psychiatrists and psychologists in Europe and North America as the fascinating book by Didier Fassin and Richard Rechtman (2009), *The Empire of Trauma*, explains in detail. Their emphasis is seeking the truth about the origins of the conception of trauma as embedded in the 'moral economy of contemporary societies' and the manner in which trauma has been created such as post-traumatic stress disorder and its politicization (Fassin and Rechtman 2009: 276). These authors ask: '... since this reality [of trauma] has only recently been recognized (that is to say, identified and legitimized), our question is: what does this social recognition change, for men and women today (whether victims or not), in their vision of the world and its history, and in their relationships with others and with themselves?' (Fassin and Rechtman 2009: 8). The answer is in what they term to be a 'politics of trauma'. This is an evocative idea but I suggest it may apply to indigenous rights and indigenous identity in a very different way than what Fassin and Rechtman have envisioned the phrase to mean through their Western heuristic approach.

While I agree with them that trauma itself is more 'a moral than ... a psychological category', and 'rather than a clinical reality, trauma today is a moral judgement' their analyses fall short in representing the 'primary agency' of social actors themselves, especially non-Western peoples in relation to Western societies (Fassin and Rechtman 2009: 284). [32] By this I mean the manner in which traumatized peoples such as Indians, whose traumatization through discrimination and racism, also instigates in them to become 'intentional beings' in using experiences of trauma as a mode of provocation against Brazilian white oppressors. What I am arguing is that stigmatizing trauma in the form of discrimination and racism actually reinforces indigenous identities. It makes them intent on overcoming defamatory remarks and malicious treatment by Brazilian white society as well as creating a counterpoint in which to establish their difference and ethnicity. The 'politics of trauma' for indigenous peoples in Manaus, because of their experiences with racism, literally becomes a point of departure in maintaining their political movements against white society. For example, in June 2009 some 300 urban Indians and some Indians from the interior occupied the federal compound of the National Health Foundation for 16 days in protest for better health care of Indians in the city and on the reservations. This demonstration marked the Indians' resolve to surmount the perceived wrongs by the non-Indian population who control the government and its agencies. Urban Amerindians therefore are not passive victims of trauma, rather their episodic memories of racism and those memories with everyday qualities of structural violence, serve as reminders as to why they must be politically active, act in concert in Pan-Amerindianism, and prevail over white Brazilian society and the government with the intent of diminishing indigenous rights (see Scheper-Hughes 1992 on everyday violence and structural violence

---

32   Here I am defining 'primary agency' from Alfred Gell (1998: 20). He states: 'primary agents, that is, intentional beings ... distribute their agency in the causal milieu, and thus render their agency effective'.

in Brazil). After all interethnic conflicts between Indians and whites have been continual for centuries in Brazil and elsewhere in Latin America, and, as in the past, Indians have resisted the maltreatment of whites, sometimes successfully but mostly ineffectively until recently.

Indigenous trauma of today stemming from racism and as experienced by Indians is not simply a moral category in which to judge Brazilian society, or the local Manauaran population, it is part of an inherited complex of past signifiers of interethnic relations over a long period of time in Brazil, and just a newer form of oppression from a multitude of historical traumatic experiences through death, genocide, ethnocide, slavery, missionization, debt peonage, land seizure, and development, and so forth. Trauma for Indians is imbued with a whole history of hate, inequalities, othering, and terror in relation to white society. Therefore, I view the politicization of trauma for Indians as literally having political consequences for them and as a means in which to address the inequalities of interethnic relations. Indigenous movements in Brazil and throughout Latin America in their current forms are the manifestations of the politicization of trauma, of overcoming inter-ethnic strife and as a means of creating dissident awareness. Indigenous protests and political movements may generally be regarded like civil rights movements as overcoming oppression with their antecedents in a distant past.

If discussing trauma is taking a moral stance then the moralities of difference must be considered. The moral economy that Fassin and Rechtman (2009) speak of has more to do with a Western, and indeed a French perspective, on the morality of trauma without much consideration for non-Westerners' moralities as juxtaposed against Western societies. What the West has conceptualized as 'trauma' and its recent acceptance in psychiatric and psychological parlance does not begin to encapsulate the terrible history of disturbing experiences of indigenous peoples in a word. It cannot. As such, trauma as just a word, and an analytical term, cannot undo the historical memories of terror for indigenous peoples wrought by interethnic conflict with white society.

Conclusions

In summary, my political engagement in advocacy for urban Amerindians in Manaus, Brazil may be characterized as more by accident than not in the sense of not knowing a priori that my fieldwork should follow some Brazilian anthropologists' leads in advocating for Indian rights. Prior to fieldwork I did not know about the indigenous politics in Manaus or for that matter not much about Indian groups living in Manaus as little is known about these city dwelling Indians outside Brazil. Also, I did not know of the many years of activism and advocacy of anthropologists in the Department of Anthropology at the Universidade Federal do Amazonas (UFAM) who had been working for the political rights of urban Amerindians and Amerindians throughout Brazil for years. I was a novice by comparison.

To many Indians I encountered I was an oddity from North America but with the possibility of being strategically important to them in their indigenous politics for reaching a wider English-speaking audience. In this manner, as with other anthropologist acquaintances, I became a 'political object' to them in a positive sense. I was someone who might prove to be politically useful to them over time. Anthropologists who become political objects to Indians serve as an example of what engagement might entail for the interethnic relationship of anthropologists with politically mindful Indians.

In my view political engagement with indigenous peoples not only entails advocacy but also evokes a multiplicity of relations between anthropologist and Indian. It implies an interethnic and intersubjective relationship, which is also reciprocal. Indians, on the one hand, use Brazilian anthropologists for their advantage in gaining access to governmental organizations, NGOs, and addressing health and economic needs and in navigating judicial issues. (In this sense, I fit in with a model of a type of a professional they were already accustomed to dealing with.) Anthropologists, on the other hand, seek empirical data in which to theorize anthropologically about Indians. Engagement is therefore a mutual dialogue and may involve indigenous peoples determining the direction of anthropological inquiry and its outcomes (as in my case). In my particular study the urban Indians of Manaus were most concerned with my promulgating their politics to a broader audience and my propagating their memories of racism and discrimination from the mistreatment of Brazilian society and Brazilian government agencies.

The reason for writing this essay is thus in support of urban Amerindians living in Manaus and to demonstrate to others the discrimination and racism they have experienced. Writings such as this chapter may add to the dedication and work of Brazilian anthropologists who have successfully advocated for the rights and recognition of urban Indians in Manaus, a continual process, and especially in negotiating with governmental agencies such as FUNAI and FUNASA.

This writing contribution however attempted to move beyond mere recognition of the Indians and their causes. Rather, it also analysed the indigenous trauma experienced through discrimination and racism and how trauma may be regarded as a catalyst for overcoming interethnic strife while at the same time defining indigeneity in juxtaposition to Brazilian society. Autobiographical knowledge and episodic memories of certain instances of racism blend with overall experiences of racism in daily life, making such negative experiences chronic or as I have argued elsewhere as forms of 'synchronic' (episodic) and 'diachronic' (reoccurring) traumas (Linstroth 2009). Therefore, these distressing episodes and disturbing structures of violence may be understood as recent traumas for indigenous peoples. Yet as I have argued here such 'trauma' has nothing to do with its conceptualization in the Western history of psychiatry and psychology but rather trauma as a mere term is only a starting point for understanding the histories of violence perpetrated by white society against Indians for more than 500 years. Racist trauma is just a modern manifestation of interethnic cruelty of whites over Indians. As I argued Fassin and Rechtman's (2009) notion of the 'politics of trauma' may be reinterpreted for indigenous peoples as those traumas, which politically instigate in Indians ways in which to maintain their political movements and resistance to white society. Another difference between Fassin and Rechtman's (2009) study and mine is an understanding that Indians (unlike other victims) are not simply victims of trauma as may be described in Western psychology. More precisely, Indians I knew have 'primary agency' as intentional beings of overcoming maltreatment. Indians also do not fit into the so-called 'moral economy' of the West where value judgements are made about the severity of trauma or whether or not it exists. Instead what must be considered are other non-Western moralities such as those of Indians judging whites and in which Indian suffering may lead to resistance. There is from a non-Western perspective the conflictive interethnic history between Indians and whites, a complex historical picture wherein oppression by whites against Indians has been the norm for centuries and long before the conceptualization of words like trauma. After all no mere words can express the history of hatred, violence, oppression, and genocide experienced by Indians in Brazil and throughout Latin America. Perhaps listening to, engaging with, and advocating for indigenous peoples and their struggles is certainly a good start albeit, in my case, a modest one.

Acknowledgements

My gratitude extends to the J. William Fulbright Foreign Scholar Grant (2008–2009) (Award #8526) and their support of my research in Manaus, Brazil in 2009. Also, I would like to thank the Universidade Federal do Amazonas (UFAM) for allowing me to be a visiting professor during the Spring Term of 2009. Most especially, I am most grateful to Raimundo Nonato Pereira da Silva and José Exequiel Basini Rodrigues whose hospitality and friendship as colleagues at UFAM I will always remember. I am also grateful to Ronald MacDonell, the priest and linguist at CIMI for his friendship and support during my fieldwork in Manaus. Additionally, I am grateful to the following people: Robin Wright of the University of Florida; Wolfgang Kapfhammer of Philipps University Marburg, and especially for Wolfgang's comments on the early stages of this essay; Clifford Brown and Michael Harris of Florida Atlantic University; and Peter Brewer of Barry University. Most of all I wish to thank my research partner, Valeria Pereira-Linstroth, for her tireless patience, support and love throughout our fieldwork period in Manaus. The data I gathered would not have been possible without her presence in the field. I am also grateful to the staff of Florida Atlantic University's Honors College, John D. MacArthur Campus Library with their tireless help in accessing resources for this chapter. Lastly, this article is also dedicated to all the urban Amerindians who befriended Valeria and me during our fieldwork together, especially the Sateré-Mawé, but to all the indigenous people we met from the ethnic groups: Apurinã, Kambeba, Kokama, Munduruku, Mura, Tikuna, and

Tukano. Words are insufficient to express our true gratitude to all the Amerindians living in Manaus who opened their doors to us and told us their stories.

References

Albert, B. 1997. Ethnographic situation and ethnic movements: Notes on post-Malinowskian fieldwork. *Critique of Anthropology*, 17(1), 53–65.

Ali, K.A. 2010. Voicing difference: Gender and civic engagement among Karachi's poor. *Current Anthropology*, 51(2), 313–20.

Asad, T. 1986. The concept of cultural translation in British Social Anthropology, in *Writing Culture: The Poetics and Politics of Ethnography*, edited by J. Clifford and G.E. Marcus. Berkeley, CA: University of California Press, 141–64.

Baer, H.A. 2012. Engaged anthropology in 2011: A view from the Antipodes in a turbulent era. *American Anthropologist*, 114(2), 217–26.

Bailey, S.R. 2009. *Legacies of Race: Identities, Attitudes, and Politics in Brazil.* Stanford: Stanford University Press.

Banks, M. 1996. *Ethnicity: Anthropological Constructions*. London: Routledge.

Bernal, R.J. 2009. *Índios Urbanos: Processo de Reconformação das Identidades Étnicas Indígenas em Manaus*. Manaus, Brazil: Universidade Federal do Amazonas, Faculdade Salesiana Dom Bosco.

Berno de Almeida, A.W. 2008a. Prefácio, in *Estigmatização & Território: Mapeamento Situacional das Comunidades e Associações Indígenas na Cidade de Manaus*, edited by A.W. Berno de Almeida and G. Sales dos Santos. Manaus, Brazil: Universidade Federal do Amazonas, 13–17.

—— 2008b. O mapeamento social os conflitos e o censo: uma apresentaçã das primeiras dificuldades, in *Estigmatização & Território: Mapeamento Situacional das Comunidades e Associações Indígenas na Cidade de Manaus*, edited by A.W. Berno de Almeida and G. Sales dos Santos. Manaus, Brazil: Universidade Federal do Amazonas, 19–33.

Berno de Almeida, A.W. and Sales dos Santos, G. (eds) 2008. *Estigmatização & Território: Mapeamento Situacional das Comunidades e Associações Indígenas na Cidade de Manaus*. Manaus, Brazil: Universidade Federal do Amazonas.

Biehl, J., Good, B. and Kleinman, A. (eds) 2007. *Subjectivity: Ethnographic Investigations*. Berkeley, CA: University of California Press.

Bloch, M.E.F. 1998. *How We Think They Think: Anthropological Approaches to Cognition, Memory, and Literacy*. Boulder, CO and Oxford: Westview Press.

Clarke, K.M. 2010. Toward a critically engaged ethnographic practice. *Current Anthropology*, 51 (2), 301–12.

Conklin, B. 1997. Body paint, feathers, and VCRs: Aesthetics and authenticity in Amazonian activism. *American Ethnologist*, 24 (4), 711–37.

Conway, M.A. 2002. Sensory-perceptual episodic memory and its context: Autobiographical memory, in *Episodic Memory: New Directions in Research*, edited by A. Baddeley, J.P. Aggleton and M.A. Conway. Oxford: Oxford University Press, 53–70.

Davis, D.J. 1999. *Avoiding the Dark: Race and the Forging of National Culture in Modern Brazil*. Aldershot, UK: Ashgate Publishing Ltd.

Dias da Costa, W. 2008. Despejos forçados de famílias indígenas, in *Estigmatização & Território: Mapeamento Situacional das Comunidades e Associações Indígenas na Cidade de Manaus*, edited by A.W. Berno de Almeida and G. Sales dos Santos. Manaus, Brazil: Universidade Federal do Amazonas, 213–28.

Dresch, P. and James, W. 2000. Introduction: Fieldwork and the passage of time, in *Anthropologists in a Wider World: Essays on Field Research*, edited by P. Dresch, W. James and D. Parkin. Oxford: Berghahn Books, 1–25.

Ewart, E. and O'Hanlon. M. (eds) 2007. *Body Arts and Modernity*. Wantage, UK: Sean Kingston Publishing.

Farmer, P. 2005. *Pathologies of Power: Health, Human Rights, and the New War on the Poor*. Berkeley: University of California Press.

Fassin, D. and Rechtman, R. 2009. *The Empire of Trauma: An Inquiry into the Condition of Victimhood* (trans. R. Gomme). Princeton and Oxford: Princeton University Press.

Ferreira de Melo, L. and Freitas Pinto, R. 2003. O migrante rural e a reconstrução da identidade e do imaginário na cidade, in *Cidade de Manaus: Visões Interdisciplinares*, edited by J. Aldemir de Oliveira, J.D. Alecrim and T.R.J. Gasnier. Manaus, Brazil: EDUA, Editora da Universidade Federal do Amazonas, 15–48.

Fixico, D.L. 2000. *The Urban Indian Experience in America*. Albuquerque, NM: University of New Mexico Press.

Foucault, M. 1980. *Power/Knowledge: Selected Interviews and Other Writings, 1972–1977*, edited by C. Gordon (trans. C. Gordon, L. Marshall, J. Mepham, and K. Soper). New York: Pantheon Books.

Gell, A. 1998. *Art and Agency: An Anthropological Theory*. Oxford: Clarendon Press.

González, N. 2010, Advocacy anthropology and education: Working through the binaries, *Current Anthropology*, 51(2), 249–58.

Graeber, D. 2001. *Toward an Anthropological Theory of Value: The False Coin of Our Own Dreams*. New York: Palgrave.

Graham, L.R. 2002. How should an Indian speak?: Amazonian Indians and the symbolic politics of language in the global public sphere, in *Indigenous Movements, Self-Representation, and the State in Latin America*, edited by K.B. Warren and J.E. Jackson. Austin, TX: University of Texas Press, 181–228.

Hale, C.R. 2006. Activist research v. cultural critique: Indigenous land rights and the contradictions of politically engaged anthropology. *Cultural Anthropology*, 2(1), 96–120.

——— 2008. Collaborative anthropologies in transition, in *A Companion to Latin American Anthropology*, edited by D. Poole. Oxford: Blackwell Publishing, 502–18.

Hanchard, M.G. 1994. *Orpheus and Power: The Movimento Negro of Rio de Janeiro and São Paulo, Brazil, 1945–1988*. Princeton: Princeton University Press.

——— (ed.) 1999. *Racial Politics in Contemporary Brazil*. Durham: Duke University Press.

Hannerz, U. 2003. Macro-scenarios: Anthropology and the debate over contemporary and future worlds. *Social Anthropology*, 11(2), 169–87.

Hastrup, K. and Elsass, P. 1990. Anthropological advocacy: A contradiction in terms? *Current Anthropology*, 31(3), 301–11.

Hemming, J. 2003. *Die If You Must: Brazilian Indians in the Twentieth Century*. London: Macmillan.

Herzfeld, M. 2010. Engagement, gentrification, and the neoliberal hijacking of history. *Current Anthropology*, 5(2), 259–67.

Howell, S. 2010. Norwegian academic anthropologists in public spaces. *Current Anthropology*, 51(2), 269–77.

Hugh-Jones, S. 1992. Yesterday's luxuries, tomorrow's necessities: Business and barter in northwest Amazonia, in *Barter, Exchange, and Value: An Anthropological Approach*, edited by C. Humphrey and S. Hugh-Jones. Cambridge: Cambridge University Press, 42–74.

Instituto Brasileiro de Geografia e Estatística (IBGE) 2010a. *Estudos & Pesquisas: Informação Demográfica e Socioeconômica* (no. 27), Síntese de Indicadores Sociais, uma Análise das Condições de Vida da População Brasileira. Rio de Janeiro, Brasil.

——— 2010b. *O Brasil Indígena*. Brasilia: Governo Federal do Brasil.

Jackson, J.E. and Warren, K.B. 2005. Indigenous movements in Latin America, 1992–2004: Controversies, ironies, new directions. *Annual Review of Anthropology*, 34, 549–73.

Jackson, J.L. 2010. On ethnographic sincerity. *Current Anthropology*, 51(2), 279–87.

Jenkins, R. 1997. *Rethinking Ethnicity: Arguments and Explorations*. London: SAGE Publications.

Johnston, B.R. 2010. Social responsibility and the anthropological citizen. *Current Anthropology*, 51(2), 235–47.

Kellett, P. 2009. Advocacy in anthropology: Active engagement or passive scholarship? *Durham Anthropology Journal*, 16(1), 22–31.

Kirsch, S. 2002. Anthropology and advocacy: A case study of the campaign against the Ok Tedi Mine. *Critique of Anthropology*, 22(2), 175–200.

Lamphere, L. 2003. The perils and prospects for an engaged anthropology: A view from the United States. *Social Anthropology*, 11(2), 153–68.

Lassiter, L.E. 2008. Moving past public anthropology and doing collaborative research. *Anthro Source*, 29, 70–86.

Linstroth, J.P. 2009. Mayan cognition, memory, and trauma. *History and Anthropology*, 20(2), 139–82.

—— 2012. The Mayan people and Sandy (Shelton) Davis: Memories of an engaged anthropologist. *Tipití: Journal of the Society for the Anthropology of Lowland South America*, 9(2), 174–209.

Low, S.M. and Merry, S.E. 2010. Engaged anthropology: Diversity and dilemmas. *Current Anthropology*, 51(2), 203–26.

Luciano, G.S. 2006. *O Índio Brasileiro: o Que Você Precisa Saber sobre os Povos Indígenas no Brasil de Hoje*. Brasília: Edições Ministério de Educação/ UNESCO.

Mahmood, C.K. 2012. A hobby no more: Anxieties of an engaged anthropology at the heart of empire. *Anthropology Today*, 28(4), 22–5.

Marx, A.W. 1998. *Making Race and Nation: A Comparison of South Africa, the United States, and Brazil*. Cambridge: Cambridge University Press.

Maximo, C.A. 2008. Mulheres Indígenas em Manaus: Conflitos sociais e burocracia na luta por um espaço politico, in *Estigmatização & Território: Mapeamento Situacional das Comunidades e Associações Indígenas na Cidade de Manaus*, edited by A.W. Berno de Almeida and G. Sales dos Santos. Manaus, Brazil: Universidade Federal do Amazonas, 95–105.

Mullins, P.R. 2011. Practicing anthropology and the politics of engagement: 2010 year in review. *American Anthropologist*, 113(2), 235–45.

Nascimento, E.L. 2007. *The Sorcery of Color: Identity, Race, and Gender in Brazil*. Philadelphia: Temple University Press.

Niezen, R. 2003. *The Origins of Indigenism: Human Rights and the Politics of Identity*. Berkeley, CA: University of California Press.

Oliveira Filho, J.P. 1998. Muita terra para pouco índio? uma introdução (crítica) ao indigenismo e à atualização do preconceito, in *A Temática Indígena Na Escola: Novos Subsídios para Professores de 1o e 2o graus*, edited by A. Lopes da Silva and L.D.B. Grupioni. São Paulo: Global Editora e Distribuidora Ltda, 61–81.

—— 1999a. A problemática dos 'índios misturados' e os limites dos estudos americanistas: um encontro entre antropologia e história, in *Ensaios em Antropologia Histórica*, edited by J.P. Oliveira Filho. Rio de Janeiro: Editora, UFRJ, 27–47.

—— 1999b. Entrando e saindo da 'mistura': os índios nos censos nacionais, in *Ensaios em Antropologia Histórica*, edited by J.P. Oliveira Filho. Rio de Janeiro: Editora UFRJ, 124–51.

——1999c. Cidadania, racismo e pluralismo: a presença das sociedades indígenas na organização do estado-nacional brasileiro, in *Ensaios em Antropologia Histórica*, edited by J.P. Oliveira Filho. Rio de Janeiro: Editora UFRJ, 192–208.

—— 2004. Uma etnologia dos 'índios misturados'?: situação colonial, territorialização e fluxos culturais, in *A Viagem da Volta: Etnicidade, Política e Reelaboração Cultural no Nordeste Indígena*, edited by J.P. Oliveira Filho. Rio de Janeiro: Contra Capa Livraria Ltda, 13–42.

Overing, J. and Passes, A. (eds) 2000. *The Anthropology of Love and Anger: The Aesthetics of Conviviality in Native Amazonia*. London: Routledge.

Ramos, A.R. 1990. Ethnology Brazilian style. *Cultural Anthropology*, 5(4), 452–72.

—— 1994. The hyperreal Indian. *Critique of Anthropology*, 14(2), 153–71.

—— 1998. *Indigenism: Ethnic Politics in Brazil*. Madison, WI: The University of Wisconsin Press.

—— 2000. Anthropologist as political actor. *Journal of Latin American Anthropology*, 4(2), 172–89.

—— 2003. Advocacy rhymes with anthropology. *Social Analysis: The International Journal of Social and Cultural Practice*, 47(1), 110–15.

—— 2008. Disengaging anthropology, in *A Companion to Latin American Anthropology*, edited by D. Poole. Oxford: Blackwell Publishing, 466–84.

—— 2012. The politics of perspectivism. *Annual Review of Anthropology*, 41, 481–94.

Restrepo, E. and Escobar, A. 2005. Other anthropologies and anthropology otherwise: Steps to a world anthropologies framework. *Critique of Anthropology*, 25(2), 99–129.

Ribeiro, D. 2000 [1995]. *The Brazilian People: The Formation and Meaning of Brazil* (trans. Gregory Rabassa). Gainesville, FL: University Press of Florida.

Sanford, V. and Angel-Ajani, A. 2006. *Engaged Observer: Anthropology, Advocacy, and Activism*. New Brunswick, NJ: Rutgers University Press.

Sansone, L. 2003. *Blackness without Ethnicity: Constructing Race in Brazil*. New York: Palgrave Macmillan.

Sawyer, L. 2009. Transforming Swedish social work with engaged anthropology. *New Proposals: Journal of Marxism and Interdisciplinary Inquiry*, 2(2), 12–17.

Scheper-Hughes, N. 1992. *Death Without Weeping: The Violence of Everyday Life in Brazil*. Berkeley: University of California Press.

—— 1995. The primacy of the ethical: Propositions for a militant anthropology. *Current Anthropology*, 36(3), 409–40.

Seigel, M. 2009. *Uneven Encounters: Making Race and Nation in Brazil and the United States*. Durham: Duke University Press.

Sheriff, R.E. 2001. *Dreaming Equality: Color, Race, and Racism in Urban Brazil*. New Brunswick, NJ: Rutgers University Press.

Sillitoe, P. 2007. Anthropologists only need apply: Challenges of applied anthropology. *Journal of the Royal Anthropological Institute*, N.S., 13, 147–65.

Smart, A. 2010. Tactful criticism in Hong Kong: The colonial past and engaging with the present. *Current Anthropology*, 51(2), 321–30.

Smith, G. 1999. *Confronting the Present: Towards a Politically Engaged Anthropology*. Oxford: Berg.

Spencer, J. 2010. The perils of engagement: A space for anthropology in the age of security? *Current Anthropology*, 51(2), 289–99.

Susser, I. 2010. The anthropologist as social critic: Working toward a more engaged anthropology. *Current Anthropology*, 51(2), 227–33.

Taussig, M. 1987. *Shamanism, Colonialism, and the Wild Man: A Study in Terror and Healing*. Chicago: University of Chicago Press.

Tomas, D. 1991. Tools of the trade: The production of ethnographic observations on the Andaman Islands, 1858–1922, in *Colonial Situations: Essays on the Contextualization of Ethnographic Knowledge*, edited by G.W. Stocking. Madison, WI: The University of Wisconsin Press, 75–108.

Turner, T. 1991. Representing, resisting, rethinking: Historical transformations of Kayapo culture and anthropological consciousness, in *Colonial Situations: Essays on the Contextualization of Ethnographic Knowledge*, edited by G.W. Stocking. Madison, WI: The University of Wisconsin Press, 285–313.

Twine, F.W. 1998. *Racism in a Racial Democracy: The Maintenance of White Supremacy in Brazil*. New Brunswick, NJ: Rutgers University Press.

Vilaça, A. and Wright, R. (eds) 2009. *Native Christians: Modes and Effects of Christianity among Indigenous Peoples of the Americas*. Surrey, UK: Ashgate Publishing Limited.

Wade, P. 2010. *Race and Ethnicity in Latin America*, 2nd edition. London and New York: Pluto Press.

Warren, J.W. 2001. *Racial Revolutions: Antiracism and Indian Resurgence in Brazil*. Durham: Duke University Press.

Wright, R. 1988. Anthropological presuppositions of indigenous advocacy. *Annual Review of Anthropology*, 17, 365–90.

Chapter 7

Old Wine in New Bottles: Self-determination, Participatory Democracy and Free, Prior and Informed Consent

Jayantha Perera

Introduction

Soon after the Second World War, the United Nations (UN) ushered in a new era in international law. Decolonization in Asia and Africa, the spread of universal concepts such as equality, justice and equity, and an emerging anti-racist consciousness (Anaya 2007) created an enabling environment to widen the scope of international law and its application. The UN Charter of 1945 embodied the key principles of international law, and one of them was the right to self-determination of peoples. Article 1(2) of the Charter outlines the purpose of international law as developing 'friendly relations among nations based on respect for the principle of equal rights and *self-determination of peoples*, and to take other appropriate measures to strengthen universal peace' (emphasis added). The right to self-determination of peoples was re-emphasized in 1966 by the UN through the International Covenant on Economic, Social and Cultural Rights (ICESCR) and the International Covenant on Civil and Political Rights (ICCPR). The common Article 1 of the two Covenants states that 'all peoples have the right of self-determination. By virtue of that right they freely determine their political status and freely pursue their economic, social and cultural development'. In addition, Article 3 of ICESCR states that 'Parties to the present Covenant … shall promote the realization of the right of self-determination, and shall respect that right, in conformity with the provisions of the Charter of the United Nations'.

Despite the presence of this enabling legal framework, states have been reluctant to use it at the sub-national level to recognize the right to self-determination of indigenous peoples who live within state boundaries. This attitude at least until the 1960s was supported by the UN because of the fear in the international community that negotiations on boundaries of peoples' territories within nation states (predominantly formed on de-colonized territories) could disintegrate such states, upsetting the fragile state-centred international political system.

During the past four decades, several indigenous movements in Latin America, North America, Australia and Asia have re-focused the attention of the UN on the right to self-determination of indigenous peoples. At the UN, their agitation led to the establishment of the Working Group on Indigenous Populations in 1982, Permanent Forum on Indigenous Issues in 2000, and the Expert Mechanism on Rights of Indigenous Peoples in 2008. Moreover, such agitations paved the path to the approval of the Indigenous and Tribal Peoples Convention, 1989 (ILO 169) and the UN Declaration on the Rights of Indigenous Peoples of 2007 (UNDRIP). Moreover, such agitations have encouraged multilateral and bilateral development agencies to recognize indigenous peoples' rights as human rights and to incorporate them in their corporate development policies and strategies. These agencies formulated corporate policies on indigenous peoples and ethnic minorities, recognizing their right to distinguish themselves from the mainstream society and to receive recognition for ancestral lands, cultural heritage, and control over natural resources.

States generally interpret the right to self-determination of indigenous peoples as a right to a better life, and opportunities for improving life chances could be found in development interventions of the state. Meanwhile, indigenous peoples interpret the scope of right to self-determination as cultural independence as a separate population which owns ancestral lands, preserves indigenous knowledge, and controls natural resources on such land. The tension between the state's determination to integrate indigenous peoples into the mainstream population, and indigenous peoples' demand to be treated as separate populations with specific rights, poses a challenge to international law. It has so far failed to resolve this tension, although some progress in indigenous peoples' participation in development and in sharing development benefits is evident. As Craven (1995) pointed out, international law arguably recognizes different needs of indigenous peoples particularly as regards their cultural identity. However the outcome of such state-sponsored engagement of indigenous peoples in development does not amount to realizing their right to self-determination guaranteed in the UN Charter.

This chapter focuses on the relationship between the right to self-determination and 'free, prior, and informed consent' (FPIC), a process that has widely been invoked to claim that indigenous peoples could enjoy the right to self-determination at least in a limited sense if they meaningfully participate in decision-making which will culminate in deciding whether to give or not their consent to a development intervention that would directly affect them. The FPIC is premised on the rights to be informed, consulted, and participate in decision-making in development projects that affect indigenous peoples, and the right to share development benefits. These rights embedded in FPIC enable indigenous peoples to demand a state to obtain their prior consent to acquire their ancestral lands or have any impact on their cultural resources, indigenous knowledge, or limit their control over natural resources. The UN Permanent Forum on Indigenous Issues, 2000 articulated FPIC as 'a process undertaken free of coercion or manipulation, involving self-selected decision-making processes undertaken with sufficient time for effective choices to be understood and made, with all relevant information provided and in an atmosphere of good faith and trust' (Tugendhat et al. 2008: 1).

The chapter examines the usefulness of FPIC in resolving the tension between state sovereignty and indigenous peoples' right to self-determination. It also looks into the role that anthropologists and indigenous scholars could play to improve FPIC as a development planning framework and its application by using their rich methodological and theoretical perspectives on indigenous communities. The chapter points out that planners of large-scale development interventions could seek advice and assistance from anthropologists and indigenous scholars to identify affected indigenous peoples of such development interventions and to study their claims over ancestral domains and cultural rights. They could also interpret domestic policies, customary laws and traditions, and record and disseminate indigenous knowledge. The Asian Development Bank recently reiterated the importance of engaging anthropologists and indigenous scholars in development interventions. 'Indigenous Peoples do not automatically benefit from development, which is often planned and implemented by those in the mainstream or dominant population. Special efforts are needed to engage Indigenous Peoples in the planning of development programs that affect them, in particular, development programs that are supposedly designed to meet their specific needs and aspirations' (2009: 55). For this purpose, the project owners or developers have to 'retain qualified and experienced experts' (ADB 2009: 58) to study the impact of such interventions on indigenous peoples, prepare plans based on detailed field work, implement development plans, and monitor and evaluate how well these plans are implemented. Anthropologists and indigenous scholars could perform these tasks.

Self-determination as an Indigenous Peoples' Right

The right to self-determination is premised on the ideal that all human beings are equally entitled to control their own destinies without external interference. It takes as its scope multiple patterns of human association and their interdependency (Xanthaki 2009). It is a collective right of a population to influence and shape decision-making processes and actions that affect its identity, economy, culture. However, during the decolonization process, under the auspices of the UN, 'the choice as to the political future of colonized peoples was not given to the individual peoples conquered, but to the inhabitants of territories colonized by European conquerors, within the boundaries of the lines of demarcation drawn by the colonizers' (Wiessner 2008: 1150). As a result, 'autonomy' of indigenous peoples in such states is an unsettled issue in international law. However, whether an indigenous population enjoys autonomy over its own affairs including ancestral land within the boundaries of a sovereign state is a political and legal question that has hitherto been decided solely by the state apparatus (Schrijver 1997; Anaya 2007; Xanthaki 2009).

ILO Convention 169 of 1989

The ILO Convention 169 focuses on the relationship between the state and indigenous peoples, and the importance of protecting indigenous peoples' decision-making rights regarding their own priorities, ancestral property, and well-being. Article 7(1) of the Convention states that they 'shall have the right to decide their own priorities for the process of development as it affects their lives, beliefs, institutions and spiritual well-being and the land they occupy or otherwise use, and to exercise control, to the extent possible, over their own economic, social and cultural development ... They shall participate in the formulation, implementation and evaluation of plans and programmes for national and regional development which may affect them directly'. This bestows on indigenous peoples some measure of control over the pace and extent of development interventions that directly affect them. By using this right, they too become accountable for the outcome of development interventions and will also have to develop strategies to benefit from them. Although the rights included in Article 7(1) seem very empowering, indigenous people can control development interventions that directly impact on them 'to the extent possible'. The phrase 'to the extent possible' creates a loophole in law and weakens the rights under the Article. The state decides the scope of this phrase and could use it to make inroads into ancestral domains of indigenous peoples without their prior consent. Such inroads not only challenges the core idea of self-determination of indigenous peoples (Sillitoe, this volume), but more importantly, can divest them of their ancestral domain and cultural heritage through involuntary resettlement, commercialization of indigenous knowledge and commercial development of natural resources (McGee 2009; ADB 2009). Moreover, although indigenous peoples have the right to participate in development programmes that affect them, they cannot veto or change them if they are likely to affect them adversely. This is because international law recognizes, and national laws emphasize that the state is responsible for regulating property and natural resources in the interests of all of its citizens.

Indigenous peoples' right to live on their customary land is recognized by Article 16(1). It states that '... the peoples concerned shall not be removed from the lands which they occupy'. But Article 16(2), (3), (4) and (5) provide exceptions to this right. Article 16(2) states 'where the relocation of these peoples is considered necessary as an exceptional measure, such relocation shall take place only with their free and informed consent'. However, if the affected indigenous peoples disagree, the state can remove them from their ancestral land, only after 'following appropriate procedures established by national laws and regulations, including public inquiries where appropriate'. A public inquiry opens up state accountability, and the participation of indigenous peoples in the enquiry legitimizes their status as the owners or at least the users of such land. But such procedures exclude the veto powers of the affected indigenous peoples, as ILO 169 recognizes state supremacy and territorial integrity. This weakens the close link recognized by international law between indigenous peoples and their ancestral land. Moreover, such state powers undermine the possibility of treating indigenous peoples as a separate sub-population within the territory of a nation state.

Indigenous peoples' right to control natural resources such as minerals on their ancestral lands is a controversial issue in many state legal systems. Article 15(1) of ILO 169 states that indigenous peoples' rights to natural resources in their ancestral land shall be 'specially safeguarded' and that they have a right to participate in decisions regarding the 'use, management and conservation' of such natural resources. But according to Article 15(2), the state can retain the ownership to mineral or sub-surface resources found in ancestral lands of indigenous peoples. The state will also prepare procedures on how to consult the peoples to ascertain whether and to what degree their interests would be prejudiced or adversely affected before undertaking or permitting any programmes for the exploration of such resources on their lands. The Article further says that 'the peoples concerned shall whenever possible participate in the benefits of such activities, and shall receive fair compensation for any damages which they may sustain as a result of such activities'. However, a state can justify the extraction of minerals or controlling of water and forest resources on such land by claiming such natural resources belong to the entire nation and should benefit all of its members. In such a situation, the probability of affected indigenous peoples receiving fair compensation is very slim.

In recent years, the increasing frequency of confrontations between indigenous peoples and multinational corporations indicate the competing ownership claims regarding natural resource management. In India, for example, the Orissa State government strongly backed the plan of a multinational corporation called Vedanta Resources plc (UK) and the Orissa Mining Corporation to extract bauxite from the Nyamgiri Hills in the State of Orissa which is home to more than 8,000 Dongaria Kondhs tribal households. The tribal community considers the hills as its ancestral domain which is sacred to them. Their agitation against the takeover of these lands for mining has been supported by the Church of England, a shareholder of the corporation, the Amnesty International and the Marlborough Ethical Fund. The Church of England sold off its shares at Vedanta Resources plc. All of them demanded that Vedanta Resources shelves its mining plans in the Nyamgiri Hills and advised the Orissa State government not to allow Vedanta Resources to mine in the hills until it obtains the tribal community's informed consent. Meanwhile, the central government of India succumbed to local and international pressure against the project and suspended it, based on the report of the Forest Advisory Committee which concluded that the mining in the proposed project area will deprive 'two Primitive Tribal Groups of their rights ... and shake the faith of tribal people in the laws of the land ...' (Saxena et al. 2010: 9). This is a rare occasion where the state succumbed to public opinion. The decision of the central government affirmed that indigenous peoples have some rights over natural resources found in their ancestral lands.

UN Declaration on Rights of Indigenous Peoples of 2007

The UNDRIP of 2007 too seeks to '... affirm the fundamental importance of the right to self-determination of all peoples, by virtue of which they freely determine their political status and freely pursue their economic, social and cultural development' (Preamble to UNDRIP). But it has not resolved the ambiguities and problems left unattended by the ILO 169. The Article 4 of UNDRIP, for example, in the spirit of the ILO 169, limits the right to indigenous peoples' right to self-determination to internal and local affairs. It states that 'Indigenous peoples, in exercising their right to self-determination, have the right to autonomy or self-government in matters relating to their internal and local affairs, as well as ways and means for financing their autonomous functions'.

Article 26 of UNDRIP states that indigenous peoples have 'the right to the lands, territories and resources which they have traditionally owned, occupied or otherwise used or acquired'. It however does not specify whether indigenous peoples have ownership rights, rather it adopts a broad approach that includes several kinds of possession including ownership. However, Article 10 comes closest to indigenous land title. It guarantees that 'no relocation shall take place without the free, prior and informed consent of the indigenous peoples concerned and after agreement on just and fair compensation and, where possible, with the option of return'. But Article 28(1) weakens and contradicts the rights guaranteed by Article 10 by introducing the possibility of acquiring or occupying of such land without obtaining their free, prior, informed consent. In such instances, as per Article 28(2), they are entitled to cash-for-land or land-for-land compensation.

Compensation for acquired land is premised on the legal principle of public purpose. It derives from the right of eminent domain of the state – the power of the state or its agent to acquire any property for a public purpose, with or without the permission of the property owner. It does not distinguish between private property and peoples' property, and the only concession available for land thus acquired is compensation, as determined by the state. In South Asia, British rulers in the nineteenth century incorporated this right of the state into country legal systems of its colonies by enacting the Land Acquisition Act of 1894. Once new nation states such as Pakistan, Ceylon (Sri Lanka), India came into being as a result of the decolonization process, each of them retained the Land Acquisition Act as the key legal instrument for acquisition of private property.

The issue of the ownership of natural resources on ancestral lands of indigenous peoples was also not resolved by UNDRIP. Article 32(2) states that the state will 'consult and cooperate in good faith with indigenous peoples in order to obtain their free, prior and informed consent before approving any project that relates to development, utilization or exploitation of mineral, water or other resources'. Compared with ILO Convention 169, UNDRIP is not a progressive piece of UN legislation. As a non-binding legal instrument it reiterates and reaffirms the powers of the state and its territorial integrity. Article 46 seals the superiority of the state vis-à-vis indigenous populations by stating that nothing in UNDRIP should be interpreted as a 'right to engage in any activity', contrary to the UN Charter or 'construed as authorizing or encouraging any action which would dismember, or impair, totally or in part the territorial integrity or political unity of sovereign and independent States'.

While developing human rights of indigenous peoples, international law continues to reiterate the state supremacy and its inherent power and right to control its territorial integrity. Thus state sovereignty can override indigenous rights, especially indigenous land and resource rights (Gilbert 2006). In this context, the inclusion of the right to self-determination in ILO Convention of 169 and in UNDRIP seems too ambitious and premature in the context of the general policy adopted by the UN and its various agencies that the right to self-determination should only be available to the whole population of a state, not to sub-populations or peoples. The degree of autonomy or self-rule of indigenous peoples variously known as 'tribal sovereignty' and 'indigenous self-government' cannot be ascertained from both ILO Convention of 169 and UNDRIP because of their lack of clarity and aspirational content.

Jurisprudence of the Right to Self-determination of Indigenous Peoples

The proposition that indigenous peoples have a right to self-determination mainly springs from the concept of aboriginal title. Aboriginal title, in a sense, insinuates a type of sovereignty held by a group of indigenous peoples over a territory before the arrival of colonizers. It is also linked with the proposition that when a colonizers claimed sovereignty over aboriginal land, such sovereignty meant the *imperium* (right to govern), but not the *dominium* (right to own) over such aboriginal land. According to this legal framework, the *dominium* over ancestral land could have been retained by the indigenous communities, unless it was expressly extinguished by a statute or by a voluntary sale or cession. Although this is a neat legal construct, its proof is difficult, and the onus rests solely on the indigenous group which challenges the state supremacy. However in recent years, courts in several countries have recognized tribal sovereignty in a limited sense, although international law is still grappling with the meaning of concepts of autonomy, self-rule, and self-determination.

In the United States v. Wheeler (1978), the Supreme Court held that tribes were 'self-governing sovereign political communities' before the arrival of the Europeans. They hold 'inherent powers of a limited sovereignty which has never been extinguished', although they no longer possess the full attributes of sovereignty. The current US legal position regarding aboriginal title and right to self-determination is well summarized in San Manuel Indian Bingo and Casino v. National Labor Relations Board (2007). In this case, the United States Court of Appeal in the District of Columbia Circuit declared that the principle of tribal sovereignty in American law exists as a matter of respect for Indian communities. In this context, the law recognizes the independence of these communities as regards internal affairs, thereby giving them the latitude to maintain traditional customs and practices. But tribal sovereignty does not amount to absolute autonomy. The court also observed that tribal sovereignty is strongest when it is based on a treaty or when the tribal authorities act within the borders of the reservation in matters concerning only tribal members.

The Australian High Court in Mabo v. Queensland (no.2, 1992) declared that the concept of *terra nullius* is historically invalid and incompatible with modern standards of human rights and justice. It recognized that common law native title had co-existed with colonial overriding title of the British Crown. However, in West Australia v. Ward (2002) and in Wilson v. Anderson (2002), the High Court confirmed that the native title may partially or totally be extinguished by competing titles granted by the Crown such as pastoral and mining leases.

In Canada, the Constitution Act of 1982 stated that 'the existing Aboriginal and treaty rights of the Aboriginal peoples of Canada are hereby recognized and affirmed'. By virtue of this provision, native rights existing at the time of the adoption of the Act, whether derived from the common law or a treaty are protected by the Constitution. The Canadian Constitution guarantees indigenous peoples' right to govern themselves regarding matters that are internal to their communities, integral to their cultures, identities, traditions, languages and institutions, and with respect to their special relationships with their ancestral lands and natural resources found on them. The Supreme Court of Canada in Sparrow v. the Queen (1990) stated that such native rights, though exist, are not absolute and could be overridden by the state. But the Sparrow judgement established the principle that 'when dealing with the rights of Aboriginal people, their rights are to be taken seriously, sensitively and in such a manner as to maintain the honour of the Crown in its fiduciary relationship with them' (Issac 1993: 214).

In Samata v. State of Andhra Pradesh (1997), the Supreme Court of India held that 'scheduled' areas (areas recognized by the Indian Constitution as where vulnerable tribal communities are predominant) should be considered as a separate category of land. Therefore those who live or depend on such lands have specific social and environmental interests in them and it is the duty of the state to respect and protect such interests. The Court also observed that 'the tribals have fundamental right to social and economic empowerment … Tribal people can exploit minerals in the "scheduled" areas without disturbing the ecology or forest lands, either individually or through cooperative societies with financial assistance from the state'. It further held that where a total prohibition on the transfer of lands does not exist, the state can allow project developers to introduce development interventions. But they must set aside at least 20 per cent of the net profits of the project as a permanent fund to meet the affected tribal people's development needs and to pay for reforestation and maintenance of the ecology.

Aspirations for self-determination and a desire to play a meaningful role in development decision-making which would affect them directly are widely shared by indigenous groups all over the world. However, they are far from being realized despite favourable judicial activism noted in several landmark judgements discussed above. The major obstacle that indigenous peoples encounter in realizing self-determination is the powerful claim of the state for sovereignty and territorial integrity. Both ILO Convention 169 and UNDRIP have succumbed to this claim. As a result, the right to self-determination of indigenous peoples is still largely limited to their right to be consulted and informed about development programmes that would affect them. Such a limited meaning of self-determination is currently couched under the label of 'participatory democracy'.

Participatory Democracy

Participatory democracy is a framework of dialogue for the state and indigenous peoples to negotiate how to resolve territorial and sovereignty issues that contaminate their relationship. It also facilitates indigenous peoples' participation in decision making and in sharing development benefits. Thus participatory democracy framework gives 'those consulted a chance to make their views known and to influence the decision' (Tomei and Swepston 1996: 42).

In the context of international law's recognition of territorial integrity and state supremacy over indigenous peoples' right to self-determination, indigenous peoples now demand state's recognition of their collective rights to ancestral domains, livelihood strategies, and cultural survival. This demand focuses on sharing and inclusion rather than on domination and exclusion. But many states still brand it as 'indigenous communalism' claiming that it clashes with a state's development strategies and interventions. They also point out that indigenous communalism would also clash with modern individualism that underpins property laws (Schrijver 1997; Anaya 2007).

The only concession that a state is generally willing to consider is to inform indigenous peoples and consult them before initiating development interventions on land where they live and earn their livelihood, but without giving them any veto power over the interventions. On the other hand, non-state actors such as NGOS and indigenous activists over the past half century have been trying to carve out more rights for indigenous peoples regarding their ancestral domains, livelihoods and cultural resources, and their control over natural resources. They also have helped the UN and international development agencies to establish procedural rights of indigenous peoples regarding consultation and disclosure of information pertaining to development interventions, initiated or supported by the state. Following the UN, several development donor agencies such as the World Bank, Asian Development Bank and International Finance Corporation have also incorporated some of these rights, especially FPIC, in varying degrees into their corporate social safeguard policies.

Development of FPIC in International Law

The Convention 169 offers guidelines as to how to apply FPIC in developing a partnership between indigenous peoples and the state in decision-making and in sharing benefits of development interventions. Articles 6.1(a) of the Convention states that governments shall 'consult the peoples through appropriate procedures and in particular through their representative institutions, whenever consideration is being given to legislative or administrative measures which may affect them directly'. Article 6(2) states that consultations 'shall be undertaken in good faith and in a form appropriate to the circumstances, with the objective of achieving agreement or consent to the proposed measures'. FPIC as a framework for dialogue among the state, indigenous peoples and development agents fits well with the 'consultation mechanism' suggested in Article 6(2). It recognizes the vulnerability of indigenous peoples, as it recognizes the need for special treatment of indigenous peoples as a separate population. It expects the state to provide full information about interventions and to allow indigenous peoples or their representatives to participate in decision-making on important policy, administrative and development issues that have direct impact them. It emphasizes that agreement or consent could be achieved only if all parties are equal in decision-making and possess full information about the issues at hand.

The ILO 169 emphasizes on 'agreement', and it does not give indigenous peoples the right to veto a state-sponsored development proposal. 'The Convention specifies that no measures should be taken against the wishes of indigenous and tribal peoples, but this does not mean that if they do not agree nothing will be done' (ILO 2003: 16). This is rather a strange statement given the fact that the Convention expects and encourages indigenous peoples or their representatives to negotiate with the state on how it could protect their rights. Negotiation implies room for change or to arrive at a conclusion acceptable to both parties. If this space is not available, the relationship becomes asymmetrical with the state having the upper hand. The state might compensate peoples for usurping their rights or establish a public inquiry into the alleged breach of customary rights. But the state could proceed with an intervention with or without the consent of the affected indigenous community. What is more revealing is that during the past 30 years, only a few states have established formal consultative procedures or frameworks to engage indigenous peoples in development interventions. As a result, consultations are generally one-sided and dominated by a state official or project officer who outlines a planned intervention to a group of vulnerable indigenous peoples expecting little or no input from them. Such consultations reduce participatory democracy to nothing. At best, they take the form of a favour from the state than a right of the affected indigenous peoples.

The UNDRIP consolidated most aspects of FPIC found in various international legal instruments. Its preamble states that '… State will apply any right it has over indigenous peoples only in consultation with them and without harming them but providing windows for opportunities for benefit sharing'. The UNDRIP attempts to balance indigenous peoples' rights and the state's duties, particularly with regard to national development. In a broad sense, its focus is more on accommodating indigenous rights than emphasizing state's sovereignty. However, as in the case of the ILO 169, it too does not bestow on indigenous peoples the right to veto a development intervention or the power to demand the state to check intervention alternatives; at best, it amounts to consultation and limited participation in decision-making. Thus FPIC continues to be a 'procedural' right rather than a 'substantive' right. As a result, it cannot be defined as a universal concept: it is too imprecise, and there is no well-established legal mechanism to enforce it in international law or in domestic legal systems.

Current Usage of FPIC in Development Interventions

The FPIC is being used in two different forms. First, it is being used as a mandatory consultation process. The World Bank uses FPIC to mean free, prior and informed *consultation*. Thus FPIC connotes consultation that 'occurs freely and voluntarily, without any external manipulation, interference, or coercion, for which the parties consulted have prior access to information on the intent and scope of the proposed project in a culturally appropriate manner, form, and language' (World Bank 2005). This means that FPIC is a process that requires an iterative series of discussions, consultations, meetings and agreement between the state and indigenous peoples who will be affected by a development intervention, supported by the World Bank.

The second form of FPIC is a consent-seeking process. The Asian Development Bank has used the key word 'consent' instead of 'consultation' in its Indigenous Peoples Policy of 1998 and also in the Safeguard Policy Statement of 2009 (ADB 1999, 2009). In the SPS, both consultation and consent are used, reserving consent to three development project-related activities, namely, commercial development of cultural resources, physical displacement of indigenous peoples, and commercial development of natural resources. The FPIC in the form of mandatory consultation is much weaker than FPIC as a consent-seeking mechanism, as 'consultation' lacks precision and as a result, is difficult to be enforced, whereas 'consent' is precise and could be ascertained.

Both 'consultation' and 'consent' emphasize the importance of meaningful consultation that establishes a qualitative measure called 'broad community support' in achieving FPIC. Thus meaningful consultation has become a process with a targeted outcome, that is, broad community support to an intervention. If there is no broad community support, the developer has to work with the community until it is achieved through good faith negotiations. However, meaningful consultation as in case of the World Bank or consent as in case of ADB needs not to be achieved fully. Both banks in their safeguard policies state that 'broad community support' could exist even if some individuals or groups object to the project activities. This is one of the weaknesses that project authorities, local governments, fieldworkers or designated officials could use to circumvent the FPIC. Cultural erosion and abject poverty in indigenous communities facilitate easy imposition of centralized decision-making that work against the FPIC process from outside. As Rosario pointed out 'most IP communities do not have long-term development plans. They live on day-to-day basis merely trying to survive. In the absence of such plans, it is hard to see how the tribes involved can really make informed decisions and ask appropriate safeguards and shared commitments' (2008: 1). In such situations, the project officials and development planners take the role of determining whether there is a 'broad community support' to the proposed development intervention without assessing the degree of community support for it.

As corporate safeguard policies on indigenous peoples indicate, multi-lateral and bi-lateral development agencies such as the World Bank (2005), ADB (2009) and International Finance Corporation (2012) and JICA (2010) have progressively been moving towards accepting that development interventions could harm indigenous peoples, their culture, rights, and therefore special action plans should be in place to protect them. For example, ADB emphasizes in its Safeguard Policy Statement that 'Indigenous peoples are increasingly threatened as development programs infringe into areas that they traditionally own, occupy, use, or view as ancestral domain' (2009: 55). International development agencies now categorically insist on indigenous peoples' agreement or consent to several specific impacts on their community, economy and culture. ADB states that 'the borrower/client will submit documentation of the engagement process to ADB for review and for ADB's own investigation to assure itself of the existence of broad community support for the project activities. ADB will not finance the project if such support does not exist' (ADB 2009). The World Bank, IFC and JICA too have similar declarations in their safeguard policies.

FPIC and Anthropology

Identification of indigenous communities that are affected by development interventions is the key to a series of subsequent activities to ensure that their rights will be secured. The formulation of comprehensive socioeconomic profiles of affected indigenous communities, screening of proposed development interventions for positive and negative impacts on them, and development of suitable consulting mechanisms among the state, development planners, and indigenous peoples are some of the key activities that facilitate the identification of affected indigenous communities, their control over natural resources and cultural systems. In this regard, international development agencies and representatives of indigenous peoples in partnership with states have developed several guidelines and best practices. These guidelines and best practices could be categorized into four key areas. The first is the formulation of a comprehensive methodology to determine whether or not a particular group of people, who are about to get affected by a development intervention, falls into the category of indigenous peoples. The second is a full legal recognition of customary land rights of indigenous peoples. The third is the recognition of the rights of indigenous peoples regarding development projects such as consultation, participation and project benefit sharing. The fourth is the prohibition of physical displacement and relocation of indigenous peoples.

Experts with wide field experience in project planning, implementation and monitoring are required to operationalize the above guidelines and best practices in development interventions. International development agencies search for such experts in academia and the NGO sector. Often they find it difficult to recruit them (Sandri, this volume). This gap could efficiently be filled by anthropologists and indigenous scholars who possess sufficient academic training in development and field experience in developing countries. They could assist governments, development agencies and NGOs in developing robust methodologies for FPIC, and in reporting ground realities – socioeconomic and political vulnerabilities of indigenous peoples. They could also recommend how international development agencies could develop their social and environmental safeguard policies to address the above four critical issues, and find better methodologies for consultation, consent-seeking and ensuring full recognition of customary land rights of indigenous peoples. A similar role could be played at the country level by those anthropologists and indigenous scholars who work at state agencies such as environmental departments, universities and NGOs.

Conclusion

The right to self-determination of indigenous peoples has to be understood within a framework which is premised on two pillars: the complete and permanent sovereignty of the state, and its territoriality. Any claim of a subpopulation within the state to ancestral land or natural resources is subject to review by the state, a tenet upheld by international law. As a result, the right to self-determination of indigenous peoples is largely confined to consultation and participation in decision-making processes which directly affect their livelihood, identity and culture, and to the right to demand that the state obtain their free, prior, and informed consent (FPIC) in case a development intervention leads to their physical displacement, exploitation of indigenous knowledge, or usurpation of their control over natural resources. However the failure to obtain their consent does not automatically result in the abandonment or termination of a proposed development intervention; the state may proceed with the intervention.

The development of FPIC as a right and a methodology for seeking consent has taken place in parallel to the growth of responsibilities of the state towards indigenous peoples. But whether FPIC has evolved to become an indigenous right itself is disputed in international law. In fact, FPIC still remains more as an aspiration and a methodology than a legal principle in international law. It is primarily an instrument that indigenous peoples or their representatives could use when they lack political and economic weight to defend or protect themselves from external interventions. Its application to other populations and communities in similar development contexts is an unsettled legal issue (McGee 2009).

Although international law has made some progress during recent decades to protect indigenous peoples, it is questionable whether international legal norms and instruments applicable to indigenous peoples' rights have been translated adequately at the national level? It is also questionable whether in effect accommodation of international legal norms and instruments in domestic legal systems has brought substantial changes to the rights and welfare of indigenous peoples? Despite the translation of some international legal instruments into national laws as in the case of the enactment of the 'Scheduled Tribes and other Traditional Forest Dwellers (Recognition of Forest Rights) Act 2006' in India, states are generally reluctant to accommodate such instruments in their domestic laws. This is because of the sensitivity of 'indigenous politics' as in case of the Chittagong Hills tribes in Bangladesh which directly challenge state sovereignty and its territorial integrity.

The possibility and desirability of elevating FPIC to the level of a right of indigenous peoples immediately confront several obstacles. One obstacle is its lack of precision. The heterogeneity of affected peoples results in attributing different meanings to FPIC. Second, generalizing 'legitimacy' of consent of given by a community to a particular intervention is difficult, because power structures of different communities will vary, and some of them could take the form of horizontal rather than hierarchical alliances, thereby limiting the effectiveness of community leaders as representatives of the community. Third, uncertainty regarding whether consensus, even if obtained, would allow a community to hand over or sell its ancestral domain to outsiders for the purposes which are not recognized in customary laws and culture. Supreme courts in several countries such as India, Philippines, and South Africa have discussed this issue at length, but there is no consensus in international law or in domestic legal systems regarding the key issue of the transferability of ancestral land to non-indigenous groups and persons.

The FPIC seeking process progressively tends to push the boundaries of indigenous peoples' right towards a form of limited self-determination. FPIC, as a general UN standard, provides an effective sphere for indigenous governance. FPIC, as a procedural law, brings some hope to indigenous peoples, particularly when state-sponsored resource exploitation and development policy ignore the legitimacy of indigenous institutions and interests. In this context, the role of FPIC is contentious, as it might be seen as a protest strategy that challenges state sovereignty, notably its control over natural resource development and management, allowing indigenous peoples to negotiate the conditions under such development and management can take place in 'their' territory (Marschke et al. 2008). This appears to run counter to the state's assertion that it has exclusive and permanent sovereignty over all natural resources in its territory, and that it has the right to use them for the benefit of all citizens of the state.

The recognition of the rights to know and to be informed enshrined in FPIC together with the right to demand that prior consent should be obtained for any development activity from indigenous peoples who might get affected from it provides the foundation for a formidable protest against a development intervention espoused by a multi-national company, as in case of Vadanta plc Group (UK), discussed above. Investors in development interventions value certainty and stability, especially in extractive industries such as coal mining and in timber logging. Such certainty and stability is absent in remote areas where mines and forests are located. These are the lands over which indigenous peoples claim their ancestral rights. Indeed investors fear agitation and strikes undermining their investments. Moreover delays arising from protests can mean huge costs to companies. In such a situation, the demand for conducting FPIC could be used by affected indigenous peoples as a strategy to buy time or discourage those who proposed the intervention. In countries where indigenous politics have reached an advanced stage, such as the Philippines, indigenous movements use FPIC in their struggle to gain legal recognition for their rights to ancestral territories. Many states therefore perceive FPIC as a 'potential' indigenous veto on national development projects.

Anthropological and indigenous scholarship can play a key role to protect the rights of indigenous peoples including their cultural rights and control over indigenous knowledge and natural resources. Many international development agencies actively seek such scholarship and expertise in identifying indigenous peoples, their ancestral domains, cultural practices and natural resources which are vital for their communal and spiritual survival. In this context, anthropologists, development planners, and international development agencies will have to work together. In this partnership, anthropologists and indigenous scholars will have to develop new methodologies which would enable development planners to scope the development impacts of interventions on indigenous communities, to check the presence or absence of broad community support for such intervention, and to prepare a strategy to ensure that they too get project benefits as planned.

References

ADB (Asian Development Bank) 1999. *Policy on Indigenous Peoples*. Manila: Asian Development Bank.

ADB (Asian Development Bank) 2009. *Safeguard Policy Statement*. Manila: Asian Development Bank.

Anaya, J.S. 2007. *Indigenous Peoples in International Law*. Oxford: Oxford University Press.

Craven, M.C.R. 1995. *The International Covenant on Economic, Social, and Cultural Rights: A Perspective on its Development*. Oxford: Clarendon Press.

Gilbert, J. 2006. *Indigenous Peoples' Land Rights Under International Law: From Victims to Actors*. New York: Transnational Publishers Inc.

Hindustan Times 2012. Retrieved from: www.hindustantimes.com (story page), accessed: 13 September 2012.

International Finance Corporation (IFC) 2012. *Performance Standard 7 – Indigenous Peoples*. Washington, DC: World Bank Group.

International Labour Office 2003. *ILO Convention on Indigenous and Tribal Peoples 1989 (No. 169): A Manual*. Geneva: International Labour Office.

Isaac, T. 1993. Balancing rights: The Supreme Court of Canada, R. v. Sparrow, and the future of Aboriginal rights. *The Canadian Journal of Native Studies*, 13(2), 199–219.

Japan International Corporation Agency (JICA) 2010. *Guidelines for Environmental and Social Considerations*. Tokyo: JICA.

Kymlicka, W. 1999. Theorizing Indigenous rights. *University of Toronto Law Journal*, 49, 281–93.

McGee, B. 2009. The community referendum: Participatory democracy and the right to free, prior and informed consent to development. *Berkeley Journal of International Law*, 27(2), 570–635.

Marschke, M., Szablowski, D. and Vandergeest, P. 2008. Engaging indigeneity in development policy. *Development Policy Review*, 26(4), 483–500.

Rosario, R. 2008. Free, Prior, and Informed Consent (FPIC): Does it give indigenous peoples more control over development of their land in the Philippines? (Unpublished M.C.P. (Master's) thesis), Department of Urban Studies and Planning, Massachusetts Institute of Technology, USA.

Saxena, N.C., Parasuraman, S., Kant, P., and Baviskar, A., 2010. Report of the Four Member Committee for investigation into the proposal submitted by the Orissa Mining Company for bauxite mining in Niyamgiri. A Report submitted to Ministry of Environment and Forests, Government of India, New Delhi on 16 August 2010.

Schrijver, N. 1997. *Sovereignty over Natural Resources: Balancing Rights and Duties*. Cambridge Studies in International Comparative Law Cambridge: Cambridge University Press.

Tomei, M. and Swepston, L. 1996. *Indigenous and Tribal Peoples: A Guide to ILO Convention No. 169*. Geneva: International Labour Organization.

Tugendhat, H., Couillard, V., Gilbert, J. and Doyle, G. 2008. *Business, Human Rights and Indigenous Peoples: The Free, Prior, Informed Consent*, a paper submitted to Joint Committee on Human Rights, House of Commons, London.

Wiessner, S. 2008. Indigenous sovereignty: A reassessment in the light of the UN Declaration on the Rights of Indigenous Peoples. *Vanderbilt Journal of Transnational Law*, 41, 1141–76.

The World Bank 2005. *Operational Policy 4.10 – Indigenous Peoples*. Washington, DC: The World Bank.

Xanthaki, A. 2009. Indigenous rights in international law over the last 10 years and future developments. *Melbourne Journal of International Law*, 10(1), 27–37.

Chapter 8

The (Non-Legal) Guide to Meaningful Recognition: A Case Study from the Canning Basin, Western Australia

Tran Tran

The laws and customs of the Karajarri, Tjurabalan and Ngurrara traditional owners have been recognized under Australian law in native title determinations over the Canning Basin aquifer. Over 10 different language groups have traditional connections to the Canning Basin area which extends from the lower Kimberley in Western Australia to the Great Sandy Desert in South Australia. Ethnographic evidence accumulated throughout the Canning Basin identifies key traditional owners with rights to 'speak for' country, with their knowledge of song and ceremony describing the importance of ancestral beings in forming and providing a template for relationships with water.

Central to the native title of the Karajarri, Tjurabalan and Ngurrara traditional owners is adherence to laws and customs based on the role of permanent freshwater springs – or *jilas. Jilas* are critical in supporting plant, animal and human life in the arid and desert lands of the Canning Basin. The consequences of native title recognition for water management are apparent in the formal and cultural role of the Karajarri, Tjurabalan and Ngurrara peoples in looking after country. As described by Karajarri woman Anna Dwyer, the concept of country is holistic and includes all living things and the interconnections between them (Dwyer 2012). In speaking about 'my country', or 'my mother's country' traditional owners assert their decision-making authority over traditional lands and waters. This authority is condensed through the native title process into legally observable laws and customs recognized according to the requirements of Section 223 of the Native Title Act 1993 (Cth) (Native Title Act) – the statutory scheme for the recognition and protection of native title under Australian property law.

Recognition of Indigenous laws and customs has been universally understood as being central to asserting Indigenous authority and achieving self-determination through political, social and economic independence (UNDRIP 2007). The Native Title Act was enacted following the Australian High Court decision of Mabo v. Queensland (No. 2) (1992) HCA 23 which recognized that Australia's Indigenous peoples continue to maintain their own robust laws and customs, and that the rights and interests flowing from these traditions are capable of protection under Australian property law. However, due to its late recognition, native title is vulnerable to extinguishment or impairment by government actions including forced removal from traditional lands and aggressive social interventions affecting Aboriginal work, home and cultural life (Strelein 2009; Martin, Bauman and Neale 2011). Native title recognition requires proof of cultural links from pre sovereignty to the present – a demanding threshold for those communities most affected by former government 'interventions' that are often cited by state government parties as proof of the cessation of cultural practice. However, despite the challenges of achieving native title recognition, 22 per cent of Australia is now covered by native title lands (National Native Title Tribunal 2013).

As the number of native title determinations throughout Australia grows, traditional owners are seeking to enforce their rights and interests against external development pressures, including water development. From the perspective of the Karajarri, Tjurabalan and Ngurrara peoples, interconnected concepts such as land use, water planning and allocation are framed as ways of looking after country. Yet these activities, while characterizing Indigenous and non-indigenous engagement with land and water, exist for governments as discrete areas of policy- making and legislation (McFarlane 2004). Within this framework, traditional owners only have a role as 'stakeholders', a position that does not equate with the responsibilities of traditional custodianship.

The Canning Basin traditional owners, and in particular the Karajarri have struggled to have their decision making authority recognized in water planning processes despite achieving the recognition of their native title lands for almost 20 years (Tran et al. 2013; Edgar 2011; Weir 2011). The protracted development of Indigenous rights and interests in water cannot be resolved without first understanding the basis of our own laws, in order to conceptualize and reconcile non-indigenous and Indigenous interests, priorities and legal traditions (Nakata 2006). That is, the struggle of the Karajarri traditional owners reflects our own inability to decolonize our thinking and enable traditional owners to better engage with and influence water allocation and use.

In this chapter I argue that Indigenous relationships to water cannot fit into the rules, laws and processes that have been imposed upon their own, without engaging with native title – not only as a legal concept but as an expression of contemporary Indigenous worldviews, cultures and identities. Engaging with these identities forms not only a symbolic element of recognition, but goes one step further towards strengthening connections back to country. This is where I fit into the story: the Karajarri, Tjurabalan and Ngurrara peoples and the Kimberley Land Council wanted to know what the impact of water planning would be on their native title and I wanted to know how traditional owners can better assert their authority and knowledge within the water allocation process (the legislation for which has been in place before native title was recognized by Australian property law). This chapter contains my reflections on why we remain constrained by the law (and the knowledge legitimated by it) and more importantly how we can limit the cumulative impact of these constraints on the Karajarri, Tjurabalan and Ngurrara traditional owners.

One Place: Many Narratives

The Canning Basin is the second largest underground aquifer in Australia, covering over 161,000 square kilometres – an area that is larger than England and Wales combined. The Basin supports irrigated agriculture, supplies community town water and is the subject of significant gas development proposals (Doust 2013; see also Paul, George and Gardiner 2013). The role of the Canning Basin in Australia's energy security often takes precedence over the role of *jilas* and fresh water flows in maintaining the health of Karajarri, Ngurrara and Tjurabalan country.

The lands of the Karajarri, Ngurrara and Tjurabalan traditional owners are interconnected via the underground water flows of the Canning Basin. Permanent fresh water plays an important role in ecological renewal and regeneration, and is linked to creative beings that inhabit the landscape. The loss of water in one area could severely compromise the health of key cultural sites located several kilometres away. Indigenous knowledges of these interconnections are based on associations between landscapes and ancestral beings that form a part of the broader knowledge economy of the region. This knowledge includes detailed understanding of the interactions between plants, animals, landscapes and people and is expressed through the laws and customs of the Karajarri, Ngurrara and Tjurabalan peoples (Barber and Jackson 2012; Tran 2013; Toussaint, Sullivan and Yu 2005). For the Canning Basin traditional owners, competition over water is not only a matter of access to a volume of water; it is also a matter of sustaining and protecting cultural and legal traditions that are linked to country.

Laws based on connections with water are also core to the Karajarri, Tjurabalan and Ngurrara people's native title determinations, hard fought and hard won by the traditional owners and their legal representatives, the Kimberley Land Council. The Karajarri are one of five groups based in the Bidyadanga Aboriginal community, formerly a Catholic mission located 200 kilometres south of Broome. The community has a population of about 800 residents consisting of the Mangala, Juwaliny, Yulparija and Nyangumarta, who were moved from their traditional lands to live on the Bidyadanga mission (Edgar 2011). The Karajarri have three separate determinations in an area covering over 33,000 square kilometres with a mix of exclusive and non-exclusive possession native title which includes the town site where the Karajarri are a minority population (Edgar 2011).

For the Karajarri, native title recognition reinstates the formal decision making power of traditional owners including the ability to control access to the land. This decision-making power, however, does not apply to 'non-exclusive' areas such pastoral lease holdings, where rights are shared with non-indigenous parties. The original Karajarri claim was driven by an agreement between the Western Australian government and Western Agricultural Industries to investigate the feasibility of large-scale cotton irrigation in the LaGrange sub-basin nestled within the Canning Basin. These proposals sparked concerns for the *jilas* on Karjarri country, leading to the lodgement and successful determination of their native title claims (Weir, Stone and Mulardy 2011; Weir 2011; Bagshaw 2003; Yu 1999).

The traditional lands of the Ngurrara and Tjurabalan peoples is located further east in the Great Sandy Desert, an area that did not experience intensive European influences, until the 1950s. The Tjurabalan and Ngurrara peoples have had limited involvement in water planning as extensive development has not occurred on their traditional lands. However the area remains under threat from potential gas development occurring in the Canning Basin that could damage the interlinked underground aquifers in the region.

The Ngurrara people are the holders of 76,000 square kilometres of exclusive possession native title recognized in Kogolo v. State of Western Australia (2007) FCA 1703. The word Ngurrara means home and the Ngurrara people's recognized rights and interests include the right to possess, occupy, use and enjoy the land to the exclusion of all others, and rights to use and enjoy the flowing and underground waters for traditional activities framed as 'hunting and fishing'. These interests have been recognized based on the Ngurrara people's connection to water which, like the Karajarri formed the key basis of their native title evidence (Vachon 2006).

Further inland, the exclusive native title rights and interests of the Tjurabalan people were recognized in Ngalpil v. Western Australia (2001) FCA 1140 over 26,000 square kilometres of land. The determination reemphasized the survival and resilience of the Tjurabalan people (at para. 23) who had occupied the area for the last 22,000 years. The determination also coincided with the establishment of the Paraku Indigenous Protected Area, which covers the inland Paruku lake system ('Lake Gregory'), a key cultural and ecological site of importance.

For the Karajarri, Tjurabalan and Ngurrara traditional owners, native title represents a turning point in their relationship with the Western Australian state government. This relationship began with an assumption that non-indigenous people were within their rights to claim the lands and waters of the Karajarri, Tjurabalan and Ngurrara people. Their traditional custodianship is based on knowledge of water places, yet early explorer diaries and publications recall the violence and coercion imposed upon Aboriginal peoples to reveal their knowledge of underground water flows. David Carnegie describes in his journal how he captured and deprived Aboriginal guides of water so as to coerce them into revealing the location of fresh water springs (Carnegie 1898: 260–61). Similarly, Alfred Canning, who sunk wells in the desert from Wiluna between 1908 and 1910, in an attempt to establish an inland stock route, commissioned chains and neck padlocks specifically for Aboriginal 'guides'. Canning's conduct resulted in a Royal Commission where he was accused of and eventually acquitted of cruelty and kidnap ('Treatment of natives: The charges against the Canning Expedition' 1908). The legacy of these attitudes continue today with important Aboriginal sites possessing European-coined names such as Godfrey Tank and Brendon Pool as opposed to *Wartikarrapungu* and *Wajanturumanu*, as they are known to the Tjurabalan traditional owners (Watson 1988). This earlier exploration also paved the way for the appropriation of Indigenous lands, facilitated by forced removals and restrictions on employment and mobility. These measures collectively supported pastoral development throughout the Kimberley, especially along the coast on Karajarri country and along the Fitzroy River (Watson 1988).

Today, Indigenous residency on native title lands is exceptionally diverse and has been influenced by these former policies. Many of the Canning Basin traditional owners continue to live in town centres such as Broome, Fitzroy Crossing and Halls Creek, as well as smaller Aboriginal communities including Bidyadanga, Jarlmadanga, Mulan, Balgo and Ringer Soak. One consistent and key priority for traditional owners after the determination of native title is to be back on country and to care for it (Strelein et al. 2013). Caring for country is a 'holistic aspect of improving health for Aboriginal people' derived from 'cultural connection to, and land management practices on Country' as well as 'sourcing and consuming natural fresh foods and identifying and gathering bush medicines for direct use to assist healing and to improve general health' (Griffiths and Kinane 2011: 25). These activities are supported by native title recognition, which has formally handed over land management responsibilities to traditional owners. The statutory scheme of the Native Title Act requires native title to be formally managed by corporate entities known as Registered Native Title Bodies Corporate (RNTBCs) (see further Bauman, Strelein and Weir 2013). These corporate structures, originally intended to facilitate non-indigenous interests over native title lands, are generally located in remote areas that typically experience high levels of poverty and limited economic development (Martin, Bauman and Neale 2011). RNTBCs seek to pursue economic development activities that are '"culturally friendly" and that support and maintain culture, language and environment' yet have limited capacity to do so (Tran, Stacey and McGrath 2013: 19).

The limited capacity of native title holders is interlinked with the broader, and oft cited disparity between Indigenous and non-indigenous socioeconomic indicators, reflecting the sustained political and social marginalization of Aboriginal and Torres Strait Islander Peoples in Australia (ABS 2008). This image of limited capacity and an inability to assert rights contrasts with the broader social demographics of the Canning Basin traditional owners. The Kimberley has one of the youngest populations in Australia with a median age of 23 years compared to 35 years for the non-indigenous population (ABS 2011). This age profile of a younger and emergent workforce necessitates the creation of opportunities to engage in community development and growth that are highly differentiated from the aging population of non-indigenous Australia. Unsurprisingly, rare and vital opportunities enabling younger generations to strengthen and maintain connections to country have become highly successful and are now emulated in existing federal government funded initiatives such as Working on Country. Working on Country reflects, in a limited way, Indigenous concepts of looking after country and provides practical resources to return to traditional lands (see further Kerins 2012).

The Karajarri and the Ngurrara peoples have established ranger programmes that are working with federal and state government funding to build water tanks and pumps on traditional lands in order to facilitate not only land management access but also the development of infrastructure to return to traditional lands.

Engagement in land and water management is a key aspiration and focus for many traditional owners and access to country correlates with the ability of Aboriginal groups to improve health and wellbeing. Accordingly, while native title has formally brought the Canning Basin traditional owners to the table they continue to struggle with the practical reality of administering vast land areas without secure funding and support that is consistent with this key aspiration (Bauman et al. 2013). Without this funding and support critical connections between staying on country and maintaining culture and identity could be seriously eroded.

Challenges to Meaningful Recognition

The legal intricacies and inadequacies of current policy and legal mechanisms for recognizing Indigenous interests in water planning and regulation has been discussed elsewhere (see for example, Tan and Jackson 2013). The consistent message that flows from this literature is that the laws of Australia habitually exclude Indigenous relationships to water; confining policy focus to agricultural production and, as a consequence, disempowers Indigenous knowledges and value systems (Weir, Stone and Mulardy 2011). There has been recent investment in research to better understand Indigenous relationships to water and what this means for water planning. However, Indigenous legal traditions (and the laws, customs and practices that inform them) will not necessarily translate into equivalent concepts in water planning (Tran 2013).

The position of the Canning Basin traditional owners is constrained by several practical and conceptual factors including: limited funding and policy support; the way in which native title is recognised by Australian law; the artificial separation of economic and cultural interests; the need for proof or evidence that can further entrench perceptions of what is and is not traditional; and the paradox created by attempting to give and retain authority at the same time.

The successful assertion of Indigenous priorities requires specific knowledge and skills to negotiate water planning processes and to understand the concept of 'allocations'. As discussed above, the Karajarri's ability to engage in formal planning processes, and thus have an opportunity to influence decision making is curtailed by a lack of funding and policy support (Bauman, Strelein and Weir 2013; Tran et al. 2013). The participation of the Karajarri is further impacted by a multitude of other pressing priorities, including negotiating housing, building economic resilience, dealing with the loss of funding for the community, poor health and poverty (Edgar 2011). Traditional owners are further excluded from the water allocation process where they are unlikely to have the funds or infrastructure to purchase or take advantage of commercial licenses. In the La Grange Basin for example, connections between people and place are not a primary factor in determining legal rights or interests (Government of Western Australia 2010a). Rather rights to water are distributed based on a commercial first come, first serve basis.

Conceptual limitations relate to how native title recognised. The Native Title Act requires Indigenous peoples to assert themselves successfully within existing regulatory frameworks. In native title, recognition of Indigenous interests is necessarily contingent on compliance with the demands of the recognition process (French 2009). Karajarri knowledge or authority is not recognized in their native title determination, only rights and interests 'to hunt, fish, gather'. In other words, Native Title law only recognizes physical activities rather than the laws and customs from which they are derived (French 2002). This process forces complex relationships to land and waters into impoverished definitions, which, when satisfied, accord traditional owners with procedural rights to protect their native title rights and interests. These procedural mechanisms require that native title holders receive notification about water extractions but do not directly enable decision making about how much and by whom water can be extracted (Government of Western Australia 2010a; see also 2010b). Notification is usually a letter or newspaper advertisement which can often be missed or not actioned for a substantial period of time. Moreover, it is unrealistic to expect a native title holder to physically collect mail (which is further compounded by the remote context in which RNTBCs operate) interpret notification letters, know the hydrogeological and cultural impact this will have, be able to present these impacts to the broader traditional owner group, and *then* write a letter in response.

Legal divisions also impact on how economic claims are dealt with. Economic interests are considered separately to cultural interests – this division is made clear by the need to associate Indigenous cultural needs within existing concepts of flow allocation from the consumptive and non-consumptive pools. For example, the water planning documents for the La Grange aquifer, assume that the provisions of the Native Title Act would protect native title and that water allocation, planning and use is separate to the recognized interests of traditional owners (Government of Western Australia 2010a). While provisions within the Native Title Act such as Section 211 provide for the recognition of cultural rights[1] any substantive Indigenous claims to water need to fit in with existing land and water management regimes, or traditional owners need to make a case for the prioritization of their interests which often involves intensive negotiations or costly litigation. The onus is on the Canning Basin traditional owners to seek resolution of Indigenous claims to water via water management legislation and policy and advocate for their interests within existing water planning processes. This requirement has two consequences: the marginalization of Indigenous legal traditions and the reaffirmation of the authority of current water planning institutions.

Where traditional owners have sought to assert their claims. the most common role of anthropology in native title has been to put Indigenous relationships between people and place before legal institutions in the form of evidence (Martin, Bauman and Neale 2011). The requirements of proof under Section 223

---

1   The recent High Court decision of Karpany v. Dietman (2013) HCA 47 affirms the broad scope for the potential application of Section 211 (para. 45–8).

of the Native Title Act, entrenches anthropology into a practice that is backward looking (Martin, Bauman and Neale 2011) despite its role in the recognition and protection of rights. For example, legal recognition demands definitive statements about claim group composition (Bauman 2006; Correy 2006). Further, the legal inquiry of native title involves the establishment of Indigenous communities as far as pre-European contact! This onerous burden of proof once satisfied, demonstrates cultural survival and vitality yet also works to entrench perceptions of what is 'traditional'. On average a native title claim takes over six years to finalise (National Native Title Tribunal 2012). Unsurprisingly, achieving native title recognition requires spending significant time collecting evidence with key senior traditional owners in very remote areas, involves recording knowledge related to water places and connecting this knowledge to family groups in order to demonstrate and successfully claim native title (McIntyre and Bagshaw 2002). This recorded knowledge is then converted into evidence in a legal claim process that can often occlude the complexities of traditional ownership. In their reflections on the Karjarri native title claim process McIntyre and Bagshaw note:

> In essence, [the Karajarri] find themselves in a position of contravening their own cultural rules, beliefs and practices by disclosing to unqualified persons their most treasured religious information for the sole purpose of obtaining basic rights to the country which they believe is, and always was, theirs. Moreover, they do so in a context in which there is a hope, but no guarantee, that the significance to them of the evidence will be understood by the Court, and that the evidence will enhance the prospects of the rights they are claiming being recognized … The difficulties, if not the very real anguish, attending this process need to be clearly understood and respected by all parties, if justice is to be adequately served. (2002: 7)

In no other context are individuals required to have their personal and cultural connections come under public scrutiny in order to establish their identities or legitimacy in exchange for what is often perceived as limited rights. While a native title claim is a legal necessity, it is also a cultural process for traditional owners who revisit and demonstrate cultural knowledge of the physical and spiritual characteristics of water based on stories, song and ceremony (Koch 2013). However, the very form of this evidence imbues Indigenous claims with mythical qualities that are not capable of encompassing explicit commercial interests (for further discussion see Strelein 2009). Within this context, relationships to water become distilled images of hunting, collecting bush foods, conducting ceremony without any connection to the identities, laws and traditions that these acts come from and sustain. As a consequence Indigenous interests are perceived as mythical or customary rather than being a proprietary interest (Sullivan, Pampila, Pajiman and Kordidi 2011).

The uniqueness of native title is also seen as a limiting factor. Many critiques of Indigenous interests in water are based on the assumption that the protection of Indigenous cultural rights (including tangible property rights) requires a form of exceptionalism, placing the recognition of Indigenous rights outside of mainstream law (Carpenter 2009). Equally there have been substantial legal and anthropological critiques of the inadequacy of property law to describe and protect Indigenous interests (Menzy 2007). Our own embedded assumptions can easily reduce rich and textured cultural relationships into seemingly mutually exclusive 'commercial' or 'cultural' interests (Carpenter 2009); and in creating greater certainty as to the precise legal nature of native title rights and interests, we curtail our ability to articulate the unique and diverse relationships that Aboriginal and Torres Strait Islander people have with their traditional lands (Strelein 2009). This paradox is replicated in water planning processes that seek to engage with native title holders and their recognized rights and interests.

The experience of the Canning Basin traditional owners reflects how debates over cultural rights are too readily narrowed into ways in which Indigenous priorities can *fit* into formal law. There are obvious problems with such an approach: Indigenous traditions, while possessing their own inherent legitimacy are forced into legalistic concepts in order to achieve recognition. Further, we end up in a position where Indigenous traditions are judged according to the formulations of our own legal institutions and processes while at the same time, paradoxically trying to pay respect to their uniqueness. Ultimately, the need to comply with the law in order to achieve recognition prevents self-determined development, by restricting decision making over water. This approach is not only contrary to the political recognition of the need for control over economic resources to overcome Indigenous disadvantage (Australian Bureau of Statistics 2008; Kimberley Land Council 2009; Steering Committee for the Review of Government Service Provision 2011), but also overlooks that recognition of Indigenous institutions which give legitimacy to the knowledge structures, concepts and existence of the peoples defined by it.

Legal arguments pursued through the native title process need to overcome distinctions made between law and custom, and tradition and commerce, distinctions that often become unclear in reality. We know these distinctions exist in the context of the Canning Basin, as existing instruments in Western Australia currently enable water extraction for mining purposes. In particular, water licences provided for mining are deemed to be valid although they are granted outside the water planning process (Tran 2013). There is similar scope to recognize Indigenous interests outside of the narrow framework of water planning. However, underlying the current limitations of recognition is a distinction made between what is and isn't Aboriginal; what is and isn't a proprietary interest; what is and isn't a legitimate form of knowledge or truth (see also Theodossopoulos this volume). If the issue is not legal, how do we navigate the conceptual challenges at hand?

Negotiating With(in) Established Institutions

The disparity between the conceptions of water planners and the priorities of traditional owners is commonly conceived of in terms of cultural difference. However, as noted above, there is a self-generating paradox that the Canning Basin traditional owners are required to navigate. The language of the law and equally its practice, attributes specialized meaning to words and concepts, constituting a specific knowledge economy and order (Dimock 2001; Moore 1973). Questions of proof or legitimacy become fundamental to the affirmation of this order which, demands a certain standard of credibility. However, the knowledge economy of native title is often limited to distinctive forms of legal and anthropological practice, which only skims the reality experienced by native title holders (Martin, Bauman and Neale 2011). The complex and intimate relationships possessed by the Karajarri, Tjurabalan and Ngurrara traditional owners cannot be fully articulated by legal processes, especially where the logic of Indigenous knowledges are excluded in the assessment of the legality or legitimacy of Indigenous relationships with country.

Greater Indigenous engagement has become a focal point in the midst of growing concerns over pressures imposed by climate change, sustained drought and the decline of the Murray Darling Basin. All these factors have attracted water planners to the seemingly vast reserves of the Canning Basin, the political framings of which, until recently, have been about resource extraction and use (Tran 2013). This increased attention has also, concurrently, drawn attention to the role and scope of Indigenous knowledges (Wallington, Robinson and Head 2012). These trends are consistent with the National Water Initiative (NWI), an agreement established in 2004 between the Australian federal and state governments in order to provide a framework for water reform.

Under the NWI water planning and allocation needs to account for native title rights and interests. State government agencies are responsible for water planning with the authority to allocate water according to established plans; creating marked variations between jurisdictions and between Indigenous groups (National Water Commission 2011). Further, state governments have argued that the law enabling native title recognition has been imposed upon them and their ability to regulate independently on matters of land and water, fuelling reluctance to 'treat native title with justice' (Strelein and Tran 2013: 48).

While aiming to be inclusive, current water law in Western Australia does not accommodate or protect Indigenous rights and interests nor does it incorporate existing Indigenous aspirations to manage and look after key water sites as a high priority. Martin, Bauman and Neale (2011: 3) argue that 'there is an increasing disjunct between the contemporary worldviews and aspirations of Aboriginal peoples and the legal construction of native title'. These contemporary world views encompass aspirations to look after country, while at the same time build community resilience and enable social development through the creation of industries on recognized native title lands.

The diverse and multiple aspirations of the Canning Basin traditional owners are confused by policy makers who are accustomed to the tone established by native title recognition which characterizes native title holders as 'traditional'. For example, the Karajarri, want to engage in their own horticultural enterprises (including reviving a successful market garden) (Edgar 2011) and argue that their entitlements to water should be considered at the outset of water planning (Tran 2013). As noted earlier, the desire of the Canning Basin traditional owners to protect their country is seen by the Western Australian state government as an environmental or cultural concern excluding them from allocations for commercial interests or forcing them to compete on a commercial scale.

The inherent contradictions of the slow decolonization of Australia's property laws are a part of the reality of the Canning Basin traditional owners. Central to the enactment of its native title regime for recognition, the Australian government sought to recognize and address the marginalization of its Indigenous peoples yet remained constrained by its own need to maintain the legitimacy of existing Australian laws and policies. We need to acknowledge connections to country as the foundation of Indigenous identities, and we need this acknowledgement to be reflected in the laws and policies that impact on the ability of the Karajarri, Tjurabalan and Ngurrara traditional owners to look after country and culture.

Current water planning is complicit in institutionalizing Indigenous knowledges within a western planning framework and assumes that pre-determined forms of participation will be sufficient for the inclusion of Indigenous priories and perspectives in decision making. However, where these institutions for decision making are dominated by a knowledge economy excluding Indigenous knowledges, it is difficult to see how participation will be sufficient to counter the policy marginalization of the Canning Basin traditional owners.

The Native Title Act requires Indigenous knowledges to have been maintained from pre-contact until the present and traditional owners have shown their strength and resilience in meeting these demands, yet they face the imposition of new standards in order to benefit from their native title. These later standards require commercial ingenuity, business knowledge and community development expertise suddenly and without adequate resourcing or support in order to maintain and protect hard won native title rights and interests. We conceive of and expect Indigenous people to live in two worlds – a misnomer in and of itself. Through our established institutions we want the Karajarri traditional owners to answer letters, read reports and assert themselves according to established standards. This approach does not make sense if we are honest about the fact that the paradoxes they face are of our making to support the authority and legitimacy of our own laws and place in the world.

Conclusion

The perceived constraints of the law to recognise Indigenous claims for both cultural and economic rights reflect established world views about the order of things or set of relationships within society. Indigenous struggles over water rights are a recurring theme in the history of the Australian nation, yet until *Mabo* had not been given serious consideration. Underlying the marginalization of the Canning Basin traditional owners from the water planning and decision-making process is non-recognition of the legitimacy of recognized native title, not only legally, but as a broader reflection of legitimate Indigenous knowledges, practices and priorities. In contrast, non-indigenous institutions, or more simply, ways of doing things, are not so easily challenged as they do not need to pass examination in order to be recognized.

This position reflects Australia's colonial inheritance where the non-recognition of Indigenous relationships to water are a part of the broader political, cultural, social and economic marginalization of First Peoples. Parallel to the emerging legal recognition of Indigenous relationships to country, recognition of the challenges of climate change, water scarcity and ecological degradation has created an opportunity for an alliance between Indigenous and non-indigenous people recognizing shared values.

However, Indigenous claims to country are seen as a threat to the integrity of current frameworks of water management. The mechanisms in place largely favour non-indigenous interests and scarcely consider the value of water in ensuring the health of communities, country and people. Indigenous claims to country are seen as just another set of interests that need to fit into the existing regime for recognition. So while there is a perception that Indigenous groups are given preferential treatment, the reality is far from this, with many Indigenous groups having minimal impact on how their country is managed.

Inherent within successful participatory processes is the ability of different parties to negotiate one another's value systems. This leads to the question of how Indigenous ways of seeing things can be better represented within water allocation and planning frameworks. Defragmenting Indigenous relationships to country requires recognition of *our own* prejudices and habits as well as the contradictions we seek to maintain through our legal institutions. This is not only a conceptual plea; there are a multitude of practical benefits that flow from investing in and recognizing the social and equity based costs of poor policy decisions that exclude Australia's First Peoples.

References

Australian Bureau of Statistics 2011. *Aboriginal and Torres Strait Islander Peoples (Indigenous) Profile*, cat. no. 2002.0. Canberra: ABS.
—— 2008. *National Aboriginal and Torres Strait Islander Social Survey*, cat. no. 4714.0. Canberra: ABS.
Bagshaw, G. 2003. *The Karajarri Claim: A Case-Study in Native Title Anthropology.* Sydney: Univeristy of Sydney (Oceania Monograph No. 53).

Barber, M. and Jackson, S. 2012. *Indigenous Water Management and Water Planning in the Upper Roper River, Northern Territory: History and Implications for Contemporary Water Planning.* Darwin: CSIRO (Report to the National Water Commission and the Department of Sustainability, Environment, Water, Population and Communities).

Bauman, T. 2006. Nations and tribes 'within': Emerging Aboriginal 'nationalisms' in Katherine. *The Australian Journal of Anthropology*, 17: 322–36.

Bauman, T., Strelein, L.M. and Weir, J.K. 2013. Navigating complexity: Living with native title, in *Living with Native Title: The Experiences of Registered Native Title Corporations*, edited by T. Bauman, L.M. Strelein and J.K. Weir. Canberra: Australian Institute of Aboriginal and Torres Strait Islander Studies (AIATSIS) Research Publications, 1–26.

Carnegie, D.W. 1898. Explorations in the interior of Western Australia. *The Geographical Journal*, 11, 258.

Carpenter, K.A. 2009. In defense of property. *Yale Law Journal*, 118: 1022–125.

Correy, S. 2006. The reconstitution of Aboriginal sociality through the identification of traditional owners in New South Wales. *The Australian Journal of Anthropology*, 17, 336–47.

Dimock, W.C. 2001. Rules of law, laws of science. *Yale Law Journal*, 13, 203.

Doust, K. 2013. Natural gas (Canning Basin joint venture) agreement bill 2013 second reading. Available at, http://www.parliament.wa.gov.au/Hansard%5Chansard.nsf/0/42143b07d64d96e848257b91000fb98c/$FILE/C39%20S1%2020130618%20p1587b-1598a.pdf. Accessed 10 December 2013.

Dwyer, A. 2012. Pukarrikarta-jangka muwarr – Stories about caring for Karajarri country. *The Australian Community Psychologist*, 24, 7–19.

Edgar, J. 2011. Indigenous land use agreement – building relationships between Karajarri traditional owners, the Bidyadanga Aboriginal Community La Grange Inc. and the government of Western Australia. *Australian Aboriginal Studies*, 2, 50.

French, R. 2002. The role of the High Court in the recognition of native title. *Western Australian Law Review*, 30, 129.

—— 2009. Native title – A constitutional shift? Paper presented at the University of Melbourne Law School, 24 March 2009.

Government of Western Australia 2010a. *La Grange Groundwater Allocation Plan* (Water Resource Allocation and Planning Series Report 25). Perth: Department of Water.

—— 2010b. *Kimberley Regional Water Plan: Supporting Detail.* Perth: Department of Water.

Griffiths, S. and Kinnane, S. 2011. *Kimberley Aboriginal Caring for Country Plan.* Broome: Nulungu Research Institute, Univeristy of Notre Dame. Available at: http://www.nd.edu.au/downloads/research/kimberley_aboriginal_caring_for_country_plan.pdf. Acessed 10 November 2013

Kerins, S. 2012. Caring for Country to Working on Country, in *People on Country, Vital Landscapes, Indigenous Futures*, edited by J.C. Altman and S. Kerins. Annandale: Federation Press, 26–44.

Kimberley Land Council 2009. Kimberley Indigenous people call for emergency summit to address disadvantage. Available at: http://uploads.klc.org. au/2013/02/090305-Call-for-Emergency-Summit.pdf. Accessed 14 October 2013.

Koch, G. 2013. *We Have the Song, So We Have the Land: Song and Ceremony as Proof of Ownership in Aboriginal and Torres Strait Islander Land Claims.* Canberra: Australian Institute of Aboriginal and Torres Strait Islander Studies (Research Discussion Paper No. 33).

McFarlane, B. 2004. The National Water Initiative and acknowledging Indigenous interests in planning. Paper delivered to the National Water Conference, Sydney, 29 November 2004.

McIntyre, G. and Bagshaw, G. 2002. Preserving culture in federal court proceedings: Gender restrictions and anthropological experts. *Land, Rights, Laws: Issues of Native Title*, 2(1), 1–12.

Martin, D., Bauman, T. and Neale, J. 2011. *Challenges for Australian Native Title Anthropology: Practice Beyond the Proof of Connection.* Canberra: Australian Institute of Aboriginal and Torres Strait Islander Studies (Research Discussion Paper No. 29).

Menzy, N. 2007. The paradoxes of cultural property. *Columbia Law Review*, 107, 2004–46.

Moore, S.F. 1973. Law and social change: The semi-autonomous social field as an approrpriate subject of study. *Law and Society Review*, 7, 719–46.

Nakata, M. 2006. Australian indigenous studies: A question of discipline. *The Australian Journal of Anthropology*, 17, 265–75.

National Native Title Tribunal 2012. *National Report: Native Title.* Perth: National Native Title Tribunal.

—— 2013. Determinations of native title. Available at: http://www.nntt.gov.au/ Mediation-and-agreement-making-services/Documents/Quarterly%20Maps/ Determinations_map.pdf. Accessed 10 December 2013.

National Water Commission 2011. *The National Water Initiative – Securing Australia's Water Future.* Canberra: National Water Initiative.

Paul, R., George, R. and Gardiner, P. 2013. *A Review of the Broome Sandstone Aquifer on the La Grange Area.* Perth: Department of Agriculture and Food (Resource management technical report no. 387).

Steering Committee for the Review of Government Service Provision 2011. Overcoming indigenous disadvantage: Key indicators. Canberra: Productivity Commission.

Strelein, L.M. 2009. *Compromised Jurisprudence: Native Title since Mabo.* Canberra: Aboriginal Studies Press.

Strelein, L. and Tran, T. 2013. Building indigenous governance from native title: Moving away from 'fitting in' to creating a decolonised space. *Review of Constitutional Studies*, 18(1), 19–48.

Strelein, L.M., Tran, T., McGrath, P.F., Powrie, R. and Stacey, C. 2013. Australian Institute of Aboriginal and Torres Strait Islander Studies (AIATSIS) submission to the Native Title Organisations Review. Canberra: Australian Institute of Aboriginal and Torres Strait Islander Studies. Available at: http://www. deloitteacceseconomics.com.au/uploads/File/AIATSIS%20Part%20B.pdf. Accessed 10 June 2014.

Sullivan, P., Pampila, H.B., Pajiman, W.B. and Kordidi, D.M. 2011. The Kalpurtu water cycle: Bringing life to the desert of the south West Kimberley, in *Country, Native Title and Ecology*, edited by J. Weir. Canberra: ANU Epress. Available at: http://press.anu.edu.au//apps/bookworm/view/country,+native+title+and+e cology/8681/ch03.html. Accessed 10 July 2013.

Tan, P.L. and Jackson, S. 2013. Impossible dreaming – Does Australia's water law and policy fulfil Indigenous aspirations? *Environmental and Planning Law Journal*, 30, 132.

Toussaint, S., Sullivan, P. and Yu, S. 2005. Water ways in Aboriginal Australia: An interconnected analysis. *Anthropological Forum*, 15, 61–74.

Tran, T. 2013. Water is country, country is culture: The translation of Indigenous relationships to water into law. PhD thesis, Univeristy of Dundee.

Tran, T., Strelein, L.M., Weir, J.K., Stacey, C. and Dwyer, A. 2013. *Native Title and Climate Change. Changes to Country and Culture, Changes to Climate: Strengthening Institutions for Indigenous Resilience and Adaptation.* Gold Coast: National Climate Change Adaptation Research Facility.

Tran, T., C. Stacey and P. McGrath 2013. Background Report on Prescribed Bodies Corporate Aspirations. In Report to Deloitte Access Economics for the FaHCSIA Review of Native Title Organisations.

UN General Assembly. *United Nations Declaration on the Rights of Indigenous Peoples: Resolution Adopted by the General Assembly*, 2 October 2007, A/RES/61/295. Available at: http://www.refworld.org/docid/471355a82.html. Accessed 12 December 2013.

Vachon, D.A. 2006. The serpent, the word and the lie of the land: The discipline of living in the Great Sandy Desert of Australia. PhD thesis, University of Toronto.

Wallington, T., Robinson, C. and Head, B. 2012. Crisis, change and water institutions in south-east Queensland: Strategies for an integrated approach, in *Risk and Social Theory in Environmental Management*, edited by T. Measham and S. Lockie. Canberra: CSIRO Publishing, 185–98.

Watson, H. 1988. Not even the cattle, in *Raparapa kularr martuwarra: All right, now we go side the river, along that sundown way: Stories from the Fitzroy River drovers*, edited by P. Marshall. Broome: Magabala Books, 106–19.

Weir, J.K. 2011. *Karajarri: A West Kimberley experience in managing native title.* Canberra: Australian Institute of Aboriginal and Torres Strait Islander Studies (Research Discussion Paper No. 30).

Weir, J.K., Stone, R. and Mulardy, M. 2011. Water planning and native title: A Karajarri and government engagement in the West Kimberley, in *Country, Native Title and Ecology*, edited by J. Weir. Canberra: ANU Epress.

Yu, S. 1999. *Ngapa Kunangkul (Living Water): Report on the Aboriginal Cultural Values of Groundwater in the La Grange Sub-Basin.* Western Australia: The Water and Rivers Commission of Western Australia.

PART III
Challenging the Dominance of the Academy

Chapter 9

Integrating African Proverbs in the Education of Young Learners: The Challenge of Knowledge Synthesis

George J. Sefa Dei
[Nana Sefa Atweneboah I]

I recognize Indigenous peoples' struggles worldwide and the imperative for students and researchers of Indigenous knowledge to show more than intellectual solidarity. I support this struggle from myriad positions as a Western-trained social anthropologist, activist educator and traditional Elder [enstooled as a Chief in Ghana]. To this end, I seek to push the edges of the intellectual envelope and consider some troublesome issues. Arguing for the validation and legitimation of Indigenous knowledges on their own terms and for us to be mindful of the limits of using the Eurocentric gaze to assess them. At the same time, I am mindful of the possibilities of Western scientific knowledge, albeit critical of its tendency to appropriate other knowledges and all the while not giving credit to, and masquerading, its understandings as universal.

This chapter contends a particular view of schooling and education in North American and African contexts. Yet, the issues that it raises are pertinent for schooling in other contexts (for example, Latin American, Caribbean, Asia and Europe). Part of a larger political and intellectual project, it re-visions schooling for youth in pluralistic contexts. It discusses the possibilities of accentuating the pedagogic and instructional role of contemporary young learners' knowledge of African Indigenous philosophies as expressed through proverbs. At the same time, I sound a caution regarding the objective of knowledge collaborations and integrations that truly allow for the engagement of multiple knowledge systems in the academy.

The chapter engages African proverbs as a form of epistemology; an Indigenous knowledge and knowing grounded on a 'long-term occupancy of a place' (i.e., the land), and derived from experiential learning and the inter-relationships of society, culture and nature. The aim in discussing the specific Ghanaian and Kenyan proverbs is to affirm such local cultural knowings as part of an Indigenous African philosophy whose teachings rests on clear associations with the land/Mother Earth and the relations of culture, society and nature. It is such teachings that inform the education of the contemporary learner when proverbs, stories, folktales, myths and mythologies are embraced in the African and other Indigenous contexts. In the discussion the chapter also brings up some contentious issues, particularly regarding the intellectual dimensions of collaborative research and the contradictions and concern studies of Indigenous ways of knowing present us.

It will be shown that African proverbs, and by extension traditional African knowledge and pedagogy, 'seek to promote values that relate to character building, imparting advice' to the young in ways contrasting 'markedly with Western knowledge and pedagogy, which seek further understanding of how the world works and how we might manage/control it, while paying little attention to producing moral persons' (Anonymous Reviewer 2013). Clearly there are differences in value systems around the world, and what comprises a moral and upright character in Africa may differ in some notable regards from European judgments. A striking example that comes through in some African proverbs is the way in which African culture stresses the notion of 'community' and judges persons by what they contribute to the collectivity or the social group to which they belong. This contrasts strongly with the way Euro-American culture promotes the 'individual' with self-interest motivating behaviour, and measuring success accordingly with a large materialistic component and as an individual accomplishment. It is opined that the West may have some important lessons to learn from African knowledge systems, given growing problems with the Western individualistic approach to life. Questions of how do we collectively build communities and create global 'communities of learners' with shared concerns about social justice, equity, love and power sharing are what makes us human. The idea that we are simply a sea of individuals has limitations in the development of communities and shared responsibilities. While the individual is a respected notion within African communities the privileged understanding is that the individual only makes sense only in relation to the community they are part of. In other words, what is valued is the 'cooperative individual' who is nourished and sustained by their communities as opposed to the 'competitive individual' who sees himself or herself and unduly over burdened by the community and shies away from collective responsibilities.

In most African communities we have proverbial teachings about community and individual, the environment and the nature: culture interface. They also cover character, probity, accountability and honesty, teachings about governance and leadership, and rights and responsibility. There are proverbial teachings about material wealth and ostentation, poverty, as well as, work ethics and division of labour. They have their instructional and pedagogic relevance for the education of youth. Proverbs help deal with everyday problems and challenges. Each proverb has its suitable occasion to be evoked, offering hope and direction for the future. When evoked inappropriately (in an improper cultural or social context) there is community disdain and sanction which may include ridicule and laughter. African traditional communities' proverbs are in high esteem given their educational purposes and teaching morals, character-building qualities as well as their understanding of local culture, tradition and history. As proverbs similar to mythologies, riddles, folktales, traditional cultural songs and stories, represent a philosophy of life, and some sanctity surrounds activities where evoked. Proverbs play a key role in the narration of such local cultural knowings. For example, many traditional stories and songs are replete with proverbs, and wise sayings that can be seen as a form of memorandum or a call to action (see Dei 2012a, b). Proverbial offers a forum of socialization around gender. It helps keep communities and peoples together.

The Culture, Schooling and Pedagogy

The culture, pedagogy and education literature, whether written from psychological, sociological and/or anthropological points of view, reinforces the relevance of the linkage of these fields (Dei 2012b, d). For example, Clifford (1986) and Geertz (1993) in anthropology demonstrate how knowledge is embedded within particular cultural contexts and resists appeals to master narratives, transcendent experiences, or a universal 'human nature'. This recognition of the specificities of knowledge is not intended to deny the shared understandings embedded in local cultural knowledges from varied communities, but it challenges claims that often masquerade as 'universal'. This appeal to universal ways of knowing, ignoring the diversity of human experiences and cultures, relies on power frameworks to privilege some forms of knowledge (for example, dominant Western ways of knowing). These scholars argue that culture informs all human experience and is central to knowledge production that offer cultures and cultural knowings are important starting points in the search for knowledge. Within psychology, Rogoff's (1981, 2003) work also details the marked differences in human development across cultures, as individuals develop as participants in their communities. This engagement is necessarily dynamic and refuses the view of culture as static and as having the same effect on each individual. The dynamism of culture is itself a product of social interactions that themselves shape understanding of culture. Such interactions may encompass the synthesis of knowledge as engaged in by and from different communities. Wertsch's (1985, 2002) studies further examine the complex relationships between individuals and culture, especially in how they take up the formulation of a collective memory, through the complex interactions between them and their cultural and historical communities. Tappan (2006) extends Wertsch's ideas (which actually can be traced back to the Enlightenment thinker) around cultural mediation to examine moral development as the accumulation of cultural tools and moral mediation means, which allow the individual to operate and make choices as part of the community. In each of these accounts, culture is the starting point for discussions on knowledge production, identity, and development (see also Cole 1992, 2006). It is the complex engagements of the individual within and with the abstraction of culture that shapes who they are and how they come to know the world.

If cultural systems constitute a way of knowing and underpin the process of coming 'to know', the challenge of knowledge integration is central, that is how we come to know and act responsibly in our communities. Ways of schooling and education in our communities are crucial. For example, how do we teach about culture, local knowledge systems (i.e. proverbs, riddles, folktales, cultural songs) to acknowledge and highlight the challenges of working with multiple knowledge? How do we as educators engage the power issues embedded in local cultural knowledge systems? How do we present education in ways that affirms local specificities and shared communities of knowledge? What is our understanding of culture? What is the link between culture and pedagogy and how do we theorize (not simply assume) this linkage? Investigating the link between culture, pedagogy, schooling and knowledge through African Indigenous philosophies, involving proverbs, folktales, riddles, stories and songs presents opportunities.

Culture is about cosmology and a worldview. Culture is also about everyday social practice. It evokes spirituality in the sense of defining what it means to be human, to understand personhood and what it is to have respect for the world of the living and the dead. The African universe is generally spiritualized and seeks to establish a communion with nature.

Cultural paradigms shape the construction of particular knowledges, as well as experiences of schooling. For example, Indigenous philosophies underline the salience of culture in producing understandings and frames of references, including what multi-centric epistemes offer. While the advancement of any one cultural perspective cannot be universally and unproblematically privileged over others, it is necessary for us to interrogate the cultural groundings of knowledge systems. In Indigenous African philosophies the idea that the elements of the universe are interrelated and intertwined (for example, mental, physical, and spiritual, material, political, economic interrelations, etc.) is entrenched. Local cultural knowledges as a body of epistemology connect place, spirit, soul and body. The wholeness of knowledge is expressed in the nexus of body, mind, spirit and soul, as well as the interrelations of culture and nature. Cultural knowledges reflect societal understandings of land, history, cultural and identity interrelations. The self and community are represented and engaged through time, history and everyday experience. Accepting the role of culture in knowledge production and making sense of our world, underscores the need to work with the notion of 'centredness' of the learner in her or his own learning in order to engage knowledge. A culturally grounded perspective that centres African/Indigenous/Asian etc. peoples' worldviews helps learners resist the dominance of hegemonic perspectives.

In schools, where the experiences and histories of marginalized groups can be denied or devalued, there is a need to centre the agency of the marginalized as learners rooted in their own histories, stories and experiences. A culture-centred paradigm provides a space for students to interpret their experiences according to their worldviews and understandings rather than being forced to do so through the dominant occidental lens. There is a lesson for the dominant too in appreciating what other cultural perspectives can offer in educating the learner. While we need to (re)construct an Indigenous outside of that identity constructed and imposed within the contexts of Euro-American hegemony that of dominant identity must also critically interrogate their values and cultural frames of reference for their omissions, negations and blind spots. African cultural values conveyed in local proverbs offer relevant philosophical, pedagogic and instructional lessons. A central lesson is the idea of a community of learners that emerge from a working knowledge of 'community belongingness' and 'the essence of social interpersonal and group relations'.

The foregoing also shows that clearly culture and schooling are linked. In conventional schooling, more often than not, the role of culture in education is misunderstood, taken as a 'problem' or a 'deficit' in educational analysis. Within the context of North American schooling, critical educators have repeatedly argued that culture and traditions are not something only certain communities possess (for example, racialized, immigrant or Indigenous and Aboriginal ones). We cannot assume any concerns about schooling are attributable to 'a problem of culture'. Working with the notion of 'problematic difference', seen through the lens of culture and difference (specifically race) is a major contributing factor to human misunderstanding. We need to understand culture (like race and other forms of difference) critically and in all its complexities. It is part of our collective human existence and we must engage with it in a positive (i.e. solution-oriented) way by paying attention to its strengths. Instead of defining culture for others, we must understand how local communities understand their culture and use such knowledge to think through contemporary challenges.

Study Context and Method

Since 2007–8, I have been involved in on-going longitudinal research in Ghana, Nigeria and Kenya examining Indigenous African philosophies (specifically knowledge systems embedded in local proverbs, songs, folktales and story forms) for their pedagogic and instructional relevance in youth education.[1] Between 2007 and 2009, over a dozen focus group discussions were organized together with workshop sessions with student-educators and teaches drawn largely from Ghana and, to some extent, Nigeria and Kenya. Over 85 individual interviews conducted with 25 educators, 20 Elders/parents, and 40 students drawn from the local universities, secondary schools and community colleges.

---

1   The initial 2007/08 study was funded through a contract grant from the Ontario Literacy and Numeracy Secretariat (LNS) for a study on 'Moral and Character Education in Ontario: Learning from African Proverbs' focusing on Ghana and Canada (see Dei 2010). The study was later extended with Social Sciences and Humanities Research Council (SSHRC) funding for a longitudinal and more comprehensive study involving Ghana, Nigeria and Kenya for the field documentation of local African proverbs and working with Canadian educators to highlight the pedagogic, instructional, communicative values in youth education.

From 2009 to 2012 continuing research has included a study of library documents and archival collections on Nigeria, Kenya and Ghanaian Indigenous cultural knowledge systems (for example parables, proverbs, riddles, songs, and folktales). In Nigeria, I worked with local undergraduate assistants, college instructors and research consultants to undertake extensive interviews with 10 Elders (as cultural custodians), twenty educators and twenty students at the Adeniran Ogunsanya College of Education in Otto/Ijanikin, and Lagos State University, Nigeria. We also conducted classroom observations at the two institutions and held a series of focus group discussions with educators and students. In the summer of 2012, we focused on basic/elementary education, vocational polytechnic training, and community-based cultural activities that contribute to cultural socialization and education of Kenyan youth. During this phase we studied the literature at the University of Nairobi and Egerton University in Egoro, Nakuru district on local proverbs, folktales, songs and other folklore. We interviewed 30 community Elders and parents, as well as 10 students at the basic/elementary school level, and 10 students at a vocational/polytechnic training institute in the Embu district. We also interviewed educators at the Kenyan institutions and conducted focus group discussions and workshops with study participants. I had trained research assistants working throughout this longitudinal study to document proverbs, riddles and folktales from Elders in local languages with English translations. The primary focus of this work has been to understand the use and meanings of local proverbs and their instructional, pedagogic and communicative values, especially teachings about identity, self-worth, respect for self, peers and authority, and the obligations and responsibilities of community belonging.

In this chapter, I focus specifically on data relating to the documentation of Ghanan and Kenyan Indigenous proverbs. It engages with the interview data on proverbs and their teachings as cultural custodians – Elders, parents, students and communities – explain them. It attempts to flesh out some of the deeply embedded meanings of these proverbs for the education of contemporary learners. In African contexts, there is a growing body of work dealing with the broad theme of 'Indigenous philosophies', including texts that document proverbs, fables, parables, tales, mythologies and their cultural meanings and interpretations (see, for example, Abubakre and Reichmuth 1997; Kalu 1991; Kudadjie 1996; Ogede 1993; Opoku 1997, 1975; Pachocinshi 1996; Yankah 1989, 1995). Many Indigenous communities elsewhere also utilize proverbs, parables, folktales and mythologies to convey knowledge on the interface of society, nature and culture (see Abrahams 1967, 1968a, 1968b, 1972; Dorson 1972; Taylor 1934; Wolfgang and Dundas 1981). We know that Aboriginal (and Maori) traditions focus more on storytelling than proverbs and fables. Yet, in Aboriginal (and Maori) epistemology we see how storytelling conveys powerful meanings similar to those encoded in proverbs and, parables in other Indigenous contexts (Firth 1926). These stories give a critical learner a powerful sense of pedagogic, instructional and communicative relevance. For example, Johnson (1993, 2003) shows Ojibway mythology as rich, complex and dense in meaning and mystery. His works provide a succinct understanding of Ojibway life, legends, and beliefs. Like stories and myths, proverbs often evoke an act of self-reflection from the learner/listener. As powerful knowledge forms, proverbs offer a deeper level of understanding and appreciation of the community's place. As the writings of Chamberlain (2003), Eastman and Nerburn (1993) and Stiffarm (1998), among many others, point out in the current globalized and transnational world where 'migration, Diaspora, and resettlement are everyday affairs' and where we continually encounter competing claims to land, resource, knowledge and power, proverbs can be helpful to all learners in appreciating our common humanity.

The question is how do we work with the teachings of such Indigenous philosophies to inform schooling for contemporary learners? Here I highlight some relevant proverbs that engage Indigenous theories and philosophies, most existing texts on African Indigenous proverbs offering brief analysis. I focus on proverbs that specifically address traditions of mutual respect, self-discipline, personal and collective identity, character building and morality, citizenship and community belonging, respect for Elders and authority, search for peace and social harmony.

Using Proverbs as Pedagogy and Instruction

It is necessary to acknowledge difficulties in translation into English and caution about situating African proverbs in Western contexts. The local contexts in which these 'wise sayings' are evoked condition their understanding. The aim is to draw on the meanings of the proverbs and what they teach youth to highlight the pedagogic and instructional role of proverbs in African Indigenous philosophies. The teachings primarily highlight individual responsibility to the community and the relevance of knowing oneself, her or his relationship to peers, and the wider community. The teachings are intended to help and guide the individual in everyday interactions whether in schools or the wider community. Adults tell youth such proverbs in their early years to instil a strong sense of self with purpose and pride.

It is should be noted that the interpretations of the proverbs offered in this chapter come from the local Elders and cultural custodians who told these problems to us as researchers. The interpretations contain far more lessons than are evident from the direct reading the proverbs alone. This is because it is generally accepted in African cultures that proverbs as wise sayings are to be interpreted in ways that meet challenges of both the teller and listener, given a social context. Consequently, the interpretations of the proverbs offered are a combination of study subjects' ideas and the analysis of what an educational researcher born within such local communities can and does make of the proverbs in light of the lessons that the tellers of these proverbs want and wish to convey to their audience/listeners.

Akan of Ghana Proverbs

> *Aboabi beka wo a na ofiri wontoma mu.*
> The insect/ant that bites you is always found in your own cloth.

This proverb teaches the young learner that while friends and those close to you may be a source of support and hope, they can also be treachery. This Akan proverb is a cautionary message. Trust is integral to betrayal with its intimate connection to relational expectations (Fitness 2001). For example, Boon defines interpersonal trust as 'the confident expectation that a partner is intrinsically motivated to take one's own best interests into account when acting – even when incentives might tempt him or her to do otherwise' (Boon 1994: 88). Clearly, trusting others exposes us to the risk them taking advantage of us. The Akan proverb warns us to be mindful of those we chose to trust because they are the ones very people who can let us down when we least expect. The lesson here is to take caution to avoid.

> *Se abofra hunu nsa hwohworo a wonne mpanyinfo didi.*
> When a child learns to wash his/her hands well, she or he eats with the Elder.

Children, in Akan tradition, are too immature to understand and appreciate the art and etiquette of eating with Elders from the same bowl. Culturally, it is an act of promotion to share the same table with Elders, particularly as in a patriarchal society. The Elders usually enjoy the finest part of a meal. Eating with Elders is an opportunity to share the best part of the meal. Metaphorically, it indicates that one's opinion is sought on matters relating to the family and community. However, this opportunity is not accorded to every child. The core test is to know how to wash one's hands. Traditionally, hand washing is not just following Western demands to meet hygienic demands. Most Akan foods are eaten with the hands, and, knowing how to wash one's hands is necessary both for one's safety and that of others. Poor hand-washing poses a danger not only to oneself but also others who share the meal. The ritual of hand washing involves soapy clean water, and towel to dry one's hands, and is done in a coordinated manner. It requires discipline, strong will and no haste. Since we eat when we are hungry, one may not be patient enough to carry out this ritual with sufficient care. If a hungry child is disciplined enough to go through this ritual successfully, s/he deserves to be treated as an adult.

Hand washing symbolizes the long term over immediate pleasure. For youngsters it relates to control of desires or temptations, the consideration of other, and ultimately betterment of society, recognizing the importance of thinking and acting collectively. Discipline, strong character and respect for oneself, peers, family, Elders and those in positions of power and authority are key aspects of maturity. Once youths demonstrate these qualities, even though young, they are invited to join the company of adults and Elders in society. Furthermore, Elders welcome youths who have shown these qualities have something to teach and contribute to society. The youth have demonstrated a character of maturity and are thus justified for being in the company of the Elders notwithstanding the junior age.

> *Abofra bo nwa na won bo akyekyedea.*
> A child cracks the shell of snails and not that of the tortoise.

In Akan tradition, children are expected to perform certain roles and adults as symbolized in this proverb. Breaking the shell of a snail is not as difficult as breaking that of a tortoise. While the snail requires little strength and less harmful tools (stones), the tortoise requires more force and dangerous tools (sharp axe). Even an adult needs patience to break the shell of a tortoise. This proverb teaches the importance of knowing one's limitations. You should act according to the expectations of your age. Humility comes with recognition of our limitations. As courage without precaution is often a step towards disaster, knowing one's limitations should not be viewed as a sign of weakness but wisdom action to ensure personal and, at times, collective safety. Youths are curious and adventurous, and should proceed with caution and acknowledge their personal limitations to avoid unwanted danger. Society establishes limitations to ensure individual and collective wellbeing.

> *Akoko sa se den ara, enye akroma fe.*
> No matter how well the fowl dances; the hawk cannot be pleased.

Enmity exists between the hawk and the fowl, such that the hawk can in no way be pleased with the actions of the fowl. The hawk must always find fault with the fowl, maintaining the tense that exists relationship between them. This proverb stresses that we cannot please everybody we meet and that there are certain persons in our lives who remind us that not all can be our friends.

According to Abraham Maslow's Hierarchy of Needs, acceptance and love are essential for human growth and development. Without these, one can feel a sense of loneliness and depression. One of the challenges of life is to gain the approval and acceptance of others. Sometimes, the desire to be accepted into a group can prompt youths act unwisely. They need to understand that not everybody will like or accept them, which is the message of this proverb. It teaches that to be successful life, they must ignore detractors and focus on in their own achievements and battle through life. The proverb concentrates on maintaining one's strength of purpose in the face of critics.

> *Benkum dware nifa na nifa so edware benkum.*
> The left hand washes the right hand and the right hand washes the left.

Reciprocity is a core pillar of Akan community. Community cannot be established and relationship building will be in danger without reciprocation. The Akan explain the essence of reciprocity through the metaphor of bathing. The left and right hands have mastered the art of reciprocity in washing are another. In this neo-liberal era of meritocracy with its cult of individualism, it is necessary that young persons are taught about the culture of reciprocity. We must teach that are cannot act with only selfish interests in mind. Instead we must recognize that we are where we are because others have learnt to carry us on their shoulders. We must acknowledge that whatever we gain in life is due to the collective efforts of others, and learn that we also have a responsibility to give back to society.

Our current approach to education is as a selfish investment. While it takes community investment makes one's education possible, the direct benefits often go to the individual. This proverb refers to the essence of community, with mutual interdependence and co-operation, responsibility and accountability, between individuals and groups. If one good turn deserves another, this means that one's actions reverberate not only on the self but on their friends, peers, and family. Contrary to the dominant conception of meritocracy, success is not an individual affair. The community contributes to any success. The lesson for schooling is in teaching that success must be inclusive of school and society. Success is not simply a question of one's academic achievements but how success contributes to the community building, social and interpersonal relations. In other words, individual success should reflect on the wider community, thus distinguishing between the 'competitive individual' who denies or shies away from the community, and the 'cooperative individual' who embraces and is simultaneously enriched by the wider community.

Yen nsan kokoromoti nho nbo epo.
One cannot bypass the thumb to tie a knot.

The thumb in this proverb represents the Creator and implies that no one can do anything without the Creator's help/guidance. The lesson here is to ask, how can one follow the path that leads to our goals in life? All learners need to know what it takes to reach set goals. We cannot neglect nor fail to pay attention to those who are instrumental in helping us do so. The thumb is also said to represent the Chief, highlighting the importance of the Indigenous institutions or respecting those in authority. The proverb teaches the importance of respecting to those from whom we can learn, and also need to recognize and appreciate those who have preceded us in our endeavours, who have helped carve the path we follow. We should let history be a guild, the wisdom and foresight of our forebears informing present and future course action.

Furthermore, this proverb teaches that a child can never outgrow her/his parents. Youths need adults to guide them through life. It also teaches that we must do first things first, taking time to ensure that all is done in a correct manner. It stresses the importance of recognizing where we are and appreciating it for what it is, being patient with ourselves to achieve our best.

Kiembu of Kenya of Proverbs

> *Kenyu na Kenyu cio canagiri nuri/nda.*
> Little kept will enrich you.

This Kenyan proverb speaks about frugality and avoiding material ostentation. It teaches that the search for what matters in life is more than material possessions. It encourages humility and appreciation of the non-material things in life such as giving, and sharing, rather than the search for material possessions. While not against material things per se, this proverb teaches a spirit of generosity, and community interdependence. It exhorts learners to be community-minded and to share knowledge and value collective help. This spirit of sharing knowledge, rather than keeping it to oneself, enriches the learning experience for all, all have something to offer. Today, competitive schooling encourages students to protect their knowledge, wanting to be recognized as intelligent and worthy of receiving the A+ and being called the 'best and brightest'. But young learners should develop a sense of interdependence and recognize that success is a collective accomplishment, not an exclusive individual one. The proverb also brings the ethic of equity, responsibility and caring for others into schooling and education. It speaks to the importance of sacrifice too: a willingness to go without for the purpose of serving others, a rare quality in our consumer driven society. It reminds us to be patient and work hard for what we want.

> *Vura mbaro ni ya ng'ondu.*
> The only good ear wax comes from the sheep.

This proverb alludes to the fact that education is the only way to come to know. But, such education is not simply school knowledge. In fact the proverbs teachers us that bright people could as well be people who might not be in the formal classroom and yet have acquired relevant knowledge through community education. The proverb does not argue schooling is not relevant but to recognize there is more to going to school that acquiring knowledge. The proverb teaches about the importance of community education and how young learners can follow the example of those who have experienced life but might not have necessarily gone to school. In effect, the educated has something to learn from the ordinary person in the street. This proverb also teaches that those who claim to be educated must first start with their own community knowledge. To acquire good education, it is necessary to advance a higher purpose of life and social existence, which rather than promoting rugged individualism, builds on sense of community responsibility and solidarity to achieve shared goals and desires. This proverb suggests that the educated is someone with vision and creativity and not necessarily formal book knowledge. Young learners need to take initiatives and become learners of their communities such that others will look up to them in society as 'educated persons'. Furthermore, the community knowledge exemplified by ordinary folks in communities can offer lessons about formal education to those in schools.

> *Murega akathwa ndaregaga agikunjiwa.*
> Always admit good advice.

In Kenyan culture when anyone, gets into trouble with the law, it is said they have failed to 'admit' to good advice in their lives. This Kiembu proverb acknowledges the importance of keeping the company of those who can advise about rights and wrongs. To admit good advice is to be disciplined, respectful, community-minded, and to appreciate and reciprocate the efforts of others. The proverb also teaches young people to learn and admit mistakes through a spirit of humility; strong persons admit to their mistakes and acknowledge that they have received 'bad advice'. One should welcome and seek out good advice from those experienced in life (for example, Elders, parents and guardians). It is said that youths who fail to heed to the advice of their parents will be taught by the outside world. You can only be advised if there is a preparedness to learn, and Elders welcome youths, who are prepared to listen to the older generation. In other words, younger persons must show a preparedness to welcome good advice to earn the companionship of Elders.

> *Kamwing ocaga ndiri/ngimi.*
> You do more when you are many.

This proverb refers the power of community over individuality. It points that we achieve more when we work collectively than individually that the 'community' is stronger than the 'individual'. It also alludes to social responsibility, no one being an island, collectives comprising a sea of individuals bounded together by shared purpose and existence. The implication is not that there are no differences or contentions, but in a community of differences, we maintain collective strength in our shared sense of belonging. We have an expectation and a responsibility to look beyond our individual desires, aspirations and concerns to uphold the social values and ethics of our communities. What an individual accomplishes can never outweigh what the community accomplishes. The proverb reminds learners who display selfish and individualistic behaviour, believing they are only responsible to themselves and their own accomplishments, that they have obligations to communities. It also teaches how individual actions have consequences for the collective. Educational success is not achieved through the individual effort of the learner alone but interaction with a community of learners. Success reflects on the community as a whole. In respect to schooling this saying teaches that educational success is achieved for all through the shared efforts of the collective (i.e. teachers, students, parents and administrators). It encourages us to move away from blaming individual learners for educational failure to analyse the collective, systemic and institutional responsibility for it.

Mwikari vakuvi na mundu mukuru ndavotagwa.
He/she who stays closer to the Elderly person will not be defeated.

This is a powerful proverb with deeply embedded teachings. One gets knowledge as wisdom. It means that he who is close to the Elderly gets a lot of knowledge and wisdom such that he is able to survive and thus will never be defeated. The proverb also exhorts the learner to seek knowledge, stressing that knowledge does not come to the idler but those who are willing to seek it. Yet, such knowledge is nothing if it does not help one to improve the world. One acquires knowledge and then act on it (for example, solve human problems). Through this process knowledge becomes wisdom, testing that learners know their identities, history, culture and sense of purpose in life. In traditional usage, this proverb distinguishes between knowledge as the acquisition of ideas and wisdom as being adept in one's society working with knowledge to transform lives. In Indigenous African philosophies, this proverb accentuates the local distinction between 'going to school' and 'receiving education' as well as the concern over 'too much schooling but too little education'. 'Going to school' does not necessarily translate is to acquiring wisdom which is the ability to apply what we learn to deal with contemporary challenges, problems and the possibilities and limitations of life. The traditional African culture considers the learner who acquires knowledge but fails in this engagement to an 'educated fool'. Pedagogically, this proverb brings a purpose and objective to education, focusing on the social relevancy in schooling. It promotes social responsibility and ethics.

Njamba ndieiwe kwao.
A good name keeps somebody.

This proverb refers to the importance of cultivating a good name for oneself, stressing how we make and create our own histories, which reflect the totality of our lived experiences and accomplishments, testimonials of living souls. One leaves a legacy leading a good and moral life worthy of being recalled as a lesson for others. To leave a footprint, the individual must have a positive impact and accomplishments that friends, peers and community will remember. This proverb teaches that one is not forgotten if one makes a good name for oneself and one's community, a testament to one's social existence. It teaches youth to lead exemplary lives, be socially responsible, respectful and support good causes, to be good character and moral, disciplined presence, respect for self, peers, Elders.

Discussion and Analysis

These teachings have lessons for all youths today notwithstanding where they live. Educators can integrate these ideas into classroom teaching alongside conventional knowledge. They may help young learners to make sense of their everyday practice and social expectations. Learners need to develop a sense of responsibility to themselves, peers, communities and society at large. It relates to the problematic of 'knowledge integration' and 'knowledge synthesis'. The synthesis of several knowledge systems is a lofty idea with possibilities (see Nakata 2007 on 'cultural interface'), but also brings challenges. What is the objective of engaging with multi-centric/multiple knowledge in educating the learner? This question relates to the mainstreaming of Indigenous knowledges. A trans-disciplinary approach to the synthesis and collaboration of multiple knowledge systems must truly rupture power relations in knowledge production. It cannot be a necessary intellectual exercise for students coming to a critical consciousness of their responsibilities as in particular Indigenous and non-Indigenous learners struggle to shed ourselves from the dominance of Western ideologies. The World Bank/IMF took a keen interest in Indigenous knowledges, in the 1990s, to further the capitalist modernist development agenda in local communities. We can question the synthesis of knowledges, if it simply serves academic and corporate capital interests. We need to ask if knowledge synthesis and collaboration is what our Indigenous communities want?

On the one hand, if we seek to maintain the currently existing parallel bodies of knowledges [not synthesis or integration], how do we challenge the dominance of Eurocentric knowledge and its tendency to devalue other bodies of thought and local communities? We may have to ensure that our work has community relevance and seek change. This is difficult in academia where we are expected to separate scholarship from our political activism. As Indigenous scholars we do not stand apart from our local communities. After all, they help sustain us in a harsh world. In this context, I want to introduce the idea of 'embodied connection' to Indigenous knowledge and trans-disciplinary research which realizes that collaborative dimensions involve more than speaking and writing in an intellectually detached way from the knowledge we produce.

What I am suggesting is that we not only discuss how our embodied knowledges speak about our social realities, but also address questions of responsibility and ethics in the use of knowledge. The survival of Indigenous communities depends on self-autonomy, self-determination, land rights, equity and justice. These issues must engage Indigenous knowledge scholarship, as members of the communities we study. We must seek to co-produce knowledge with our communities in ways that fundamentally shift established ways of knowledge production.

There are limits to intercultural dialogue. My work on inclusion in Canadian contexts has taught me to be critical of inter-cultural dialogue as a gateway to social inclusion. We need to trouble the bland talk of 'inclusion/diversity' with pointed critiques of power, accountability and transparency. We must begin by asking: inclusion into what? How do Indigenous knowledges and mainstream knowledges engage in 'dialogue' in an academy with a 'spiritual proof fence' (Shahjahan 2007; see also Masseri 1994) that is denied spirituality as 'anti-intellectual'? We know that many of our academic colleagues shiver when we say we are having an intellectual conversation with the 'spirit/spiritual', with emotions and intuitions! In searching for 'multi-epistemic spaces' (Cajete 2000) and becoming 'border-crossers' (Mignolo 2002) we must search for ways to validate Indigenous knowledges in their own right and not through terms of engagement set up by dominant Western scholarship.

Elsewhere, I have advanced the notion of creating (*Suahunu*) a 'trialectic space in the academy to structure dialogue' (Dei 2012c). '*Suahunu*' is an Akan concept stressing coming to learn and know and to act responsibly. The '*Suahunu*' trialectic space involves a dialogue between multiple parties, a sort of 'dialogic encounters' with an epistemic community, a community of shared intellectual leaning and political praxis. It is constituted as a space where learners can openly engage the body, mind and spirit/soul interface in critical dialogues about their education. It is also a space that nurtures conversations that acknowledge the importance and implications of working with a knowledge base about society, culture, and nature nexus. Such spaces can only be created when we open our minds broadly to envision schooling as an opportunity to challenge dominant paradigms and academic reasoning.

Another issue of regarding power relations and dynamics is the 'incommensurability of certain knowledges', which relates to philosophical differences between them. These pose some dilemmas: as Andreotti, Ahenakew and Cooper (2011) argue, if Indigenous scholars present Indigenous knowings as 'too different' they risk being interpreted as making 'no sense' and therefore not counting as 'knowledge'. Or, if they present their knowings as similar to dominant ways of knowing, then they may be perceived as adding nothing new to Eurocentric knowledge.

The search for 'epistemological pluralism' has to acknowledge some fundamental tensions, and contradictions in translating Indigenous knowledges and epistemologies into Western, i.e. non-Indigenous languages, categories and technologies. For example, how do we as Indigenous scholars use the colonial language (in this case English), Western 'terminologies and logics (for example epistemology, ontology, axiology) and technologies (alphabetical writing, digital scripts)' to address issues relating to Indigenous ways of knowing (Andreotti, Ahenakew and Cooper 2011: 44)? There is an inherent paradox in Indigenous scholars being obliged to communicate about Indigenous knowledge in ways that are 'intelligible and coherent … to readers and interpreters in the dominant culture' (Andreotti, Ahenakew and Cooper 2011: 45). Also, what Indigenous scholars are to do when their translations of their ways of knowing as are perceived by our own communities as a 'perversion, corruption or [even] unfair appropriation of Indigenous epistemologies' (Andreotti, Ahenakew and Cooper 2011: 45)? The 'politics of knowledge synthesis' needs to address there is the difficulty of working with Indigenous ways of knowing that do not fit the 'parameters of acceptability' established by so-called modern knowledge (see also Andreotti, Ahenakew and Cooper 2011: 42; Santos 2002).

I end with this caution. The concerns raised above should not stop us from the task at hand, working for knowledge collaborations, integration and perhaps synthesis. But we need to exercise caution. The Akan of Ghana say: 'The one who has been bitten by a snake is always afraid of the worm'. Similarly, the Yoruba of Nigeria saying: 'It is the adamant fly that follows the corpse to its grave'.

To reiterate this chapter has sought to present a view on Indigenous knowledge scholarship using African proverbs. The central argument is African knowledge and traditional teaching stress the imparting of values that make for good character, as well as knowledge and education that focuses on furthering understanding of our worlds. However, there is a question of difference in the core social values privileged between cultures. For example, the emphasis as to what makes a good and moral character is different in different contexts. The chapter makes a contrast between the community values of African cultures that highlight the interests in the collectivity and those of the West that privilege individual self-interest. Arguably, the West and learners in the current Euro-American school system have much to learn from African philosophies of education in this respect, as the individualistic focus of the West has serious limitations in building healthy, sustainable schooling communities with a 'community of learners'.

Acknowledgements

I have so many Ghanaian, Nigerian, Kenyan and Canadian local research assistants and consultants, students, parents and Elders to thank and I am afraid I will miss some very important helpers as well. In Nigeria there is Lateef Layiwola, Joy Odewumi, Chinyere Eze, Provost Hakeem Olato Kunbo Ajose-Adeogun, Tola Olajuwon, Dr A O.K. Noah, not to mention the many students and educators at the Adeniran Ogunsanya College of Education in Otto/Ijanikin, Lagos State, and the Lagos State University, Lagos. In Kenya I want to thank particularly Samuel Njagi, Grace Makumi, Moodley Phylis, and the students and educators at Eggerton University, Ngoro, Nakuru and the University of Nairobi, Gichugu Primary School, Kandori Youth Polytechnic in the Embu area, Kenya. In Ghana special thanks to Anane Boamah, Osei Poku, Kate Araba Stevens, Daniel Ampaw, Ebenezer Aggrey, Paa Nii, Alfred Agyarko, Professor Kola Raheem, and the many students, educators at local universities and parents and Elders who generously gave their time and expertise to ensure the success of the field study. At the University of Toronto in Canada I am grateful to Dr Paul Adjei, Dr Lindsay Kerr, Harriet Akanmori, Jennifer Jaguire, Isaac Darko, Yumiko Kawano, Jadie McDonnell, Dr Bathseba Opini, Shaista Patel, Mini Tharakkal, and Michael Nwalutu. I am particularly thankful to Yumiko Kawano for her continued assistance in responding to reviewers' and editorial comments. Finally I want to thank the Social Science Research Council of Canada for continued funding of this project.

References

Abrahams, R. 1967. On proverb collecting and proverb collection. *Proverbium*, 8, 181–4.

—— 1968a. A rhetoric of everyday life: Traditional conversational genres. *Southern Folklore Quarterly*, 32, 44–59.

—— 1968b. Introductory remarks in a rhetorical theory of folklore. *Journal of American Folklore*, 81, 143–58.

—— 1972. Proverbs and proverbial expression, in *Folklore and Folklife*, edited by R. Dorson. Chicago: Chicago University Press, 117–27.

Abubakre, R.D. and Reichmuth, S. 1997. Arabic writing between global and local culture: Scholars and poets in Yorubaland. *Research in African Literatures*, 28(3), 183–209.

Andreotti, V, C. Ahenakew and G. Cooper (2011). "Epistemological Pluralism: Ethical and Pedagogical Challenges in Higher Education". *AlterNative: An International Journal of Indigenous Peoples,* vol.7, no.1.

Anonymous Reviewer 2013. Comments submitted to the author from a reviewer of the paper. Janaury 2013.

Boon, S. 1994. Dispelling doubt and uncertainty: Trust in romantic relationships, in *Dynamics of Relationships: Understanding Relationship Processes Vol. 4*, edited by S. Duck. Thousand Oaks: Sage, 86–111.

Cajete, G. 2000. *Native Science: Natural Laws of Interdependence.* Santa Fe: Clear Light Publishers.

Chamberlain, T. 2003. *If this is your Land, Where are your Stories?: Finding Common Ground.* Toronto: Alfred A. Knopf Canada.

Clifford, J. 1986. On ethnographic allegory, in *The Poetic and Politics of Ethnography: Experiments in Contemporary Anthropology*, edited by J. Clifford and G.E. Marcus. California: University of California Press, 98–121.

Cole, M. 1992. Context, modularity, and the cultural constitution of development, in *Children's Development within Social Context, Vol. 2: Research and Methodology*, edited by L.T. Winegar and J. Valsiner. Hillsdale: Lawrence Erlbaum Associates, 5–31.

—— 2006. *The Fifth Dimension: An After-school Program Built on Diversity.* New York: Russell Sage.

Dei, G.J.S. 2010. *Reclaiming Indigenous Knowledge through Character Education: Implications for Addressing and Preventing Youth Violence.* A Final Report submitted to the Literacy and Numeracy Secretariat (LNS), Ministry of Education, Ontario. 1 February 2010. (With the assistance of Jagjeet Gill, Camille Logan, Dr Meredith Lordan, Marlon Simmons and Lindsay Kerr.)

—— 2012a. Learning form Indigenous cultural stories: The case of Ananse stories. Paper presented on panel session, *Trickster Stories and Other Stories from Indigenous Communities: Relevance for Education*, organized by

Judy Iseke Barnes for the Annual Meeting of the American Educational Research Association (AERA), Vancouver, BC, April, 13–17.

—— 2012b. 'Interrogating democratic education' as global citizenship education: Thinking out differently. Paper presented on the panel session on *Re-Imagining the 'Global Citizenship' Discourse: A View from the Contexts of Developing Countries.* Organized for the Annual Meeting of the American Educational Research Association (AERA), Vancouver, BC, April 13–17.

—— 2012c. 'Suahunu': The trialectic space. *Journal of Black Studies*, 43(8), 823–46.

—— 2012d. Indigenous anti-colonial knowledge as 'heritage knowledge' for promoting black/African education in diasporic contexts. *Decolonization: Indigeneity, Education and Society*, 1(1), 1–18.

Dorson, R. (ed.) 1972. *Folklore and Folklife.* Chicago: Chicago University Press.

Durkheim, E. 1897. *Suicide.* Beverly Hills, CA: Sage Publication.

Eastman, C.A. and Nerburn, K. (eds) 1993. *The Soul of an Indian and other Writings from Ohiyesa (The Classice Wisdom Collection).* California: New World Library.

Firth, R. 1926. Proverbs in the native life with particular reference to those of the Maori. *Folklore*, 37, 134–53.

Fitness, J. 2001. Betrayal, rejection, revenge, and forgiveness: An interpersonal script approach, in *Interpersonal Rejection*, edited by M. Leary. New York: Oxford University Press, 73–103.

Gaylin, W. 1984. *The Rage Within: Anger in Modern Life.* New York: Simon and Schuster.

Geertz, C. 1993. *Local Knowledge: Further Essay in Interpretive Anthropology.* London: Fontana Press.

Johnson, B. 1993. *Ojibway Tales.* Nebraska: University of Nebraska Press.

—— 2003. *Ojibway Heritage.* Toronto: McClelland and Stewart.

Jones, W.H., Couch, L. and Scott, S. 1997. Trust and betrayal: The psychology of getting along and getting ahead, in *Handbook of Personality Psychology*, edited by R. Hogan, J. Johnson and S. Briggs. New York: Academic Press, 465–82.

Kalu, Ogbu U. 1991. Gender ideology in Igbo religion: The changing religious role of women in Igboland. *Africa/Istituto Italo-Africano*, 46(2), 184–202.

Kudadjie, J.N. 1996. *Ga and Dangme Proverbs: For Preaching and Teaching.* Accra: Asempa Publishers. Available at: www.afriprov.org/ebooks/gadangme. htm, accessed: 20 May 2005.

Masseri, A. 1994. The imperialist epistemological vision. *The American Journal of Islamic Social Science*, 11(3), 403–15.

Maxwell, J.C. 1993. *The Winning Attitude.* Nashville: Thomas Nelson Publishers.

Mignolo, W. 2000. The geopolitics of knowledge and the colonial difference. *The South Atlantic Quarterly*, 101(1), 57–94.

Nakata, M. 2007. *Disciplining the Savages: Savaging the Disciplines.* Canberra, ACT: Aboriginal Studies Press.

Ogede, Ode S. 1993. The role of the Igede poet Micah Ichegbeh's 'Adiyah' songs in the political and moral education of his local audiences. *African Languages and Cultures*, 6(1), 49–68.

Opoku, K.A. 1975. *Speak to the Winds: Proverbs from Africa.* New York: Northrop, Lee and Shepard Co.

—— 1997. *Hearing and Keeping. Akan Proverbs*. Accra: Asempa Publishers.

Pachocinshi, R. 1996. *Proverbs of Africa: Human Nature in the Nigerian Oral Tradition: An Exposition and Analysis of 2,600 Proverbs from 64 Peoples.* Continuum International Publishing. Available at: http://www.paragonhouse. com/catalog/product_info.php?cPath=23_46andproducts_id=155, accessed: October 2014.

Rogoff, B. 1981. Schooling and the development of cognitive skills, in *Handbook of Cross-cultural Psychology, Vol 4: Developmental Psychology*, edited by H.C. Triandis and A. Heron. Rockleigh: Allyn and Bacon, 233–94.

—— 2003. *The Cultural Nature of Human Development*. New York: Oxford University Press.

Santos, B. 2002. Toward a multicultural conception of human rights, in *Moral Imperialism: A Critical Anthology*, edited by B. Hernandez-Truyol. New York: New York University Press, 39–60.

Shahjahan, R.A. 2007. The everyday as sacred: Trailing back by the spiritual proof fence in the academy. Unpublished doctoral dissertation, Ontario Institute for Studies in Education, University of Toronto, Toronto.

Stiffarm, L.A. (ed.) 1998. *As We See ... Aboriginal Pedagogy*. Saskatoon: University Extension Press. University of Saskatchewan.

Tappan, M.B. 2006. Moral functioning as mediated action. *Journal of Moral Education*, 35(1), 1–18.

Taylor, A. 1934. Problems in the study of proverbs. *Journal of American Folklore*, 47, 1–21.

Wertsch, J.V. 1985. *Vygotsky and the Social Transformation of Mind.* Cambridge: Harvard University Press.

—— 2002. *Voices of Collective Remembering*. Cambridge: Cambridge University Press.

Wolfgang, M. and Dundas, A. 1981. *The Wisdom of Many: Essays on the Proverb.* New York: Garland Publishing

Yankah, K. 1989. *The Proverb in the Content of Akan Rhetoric: A Theory Proverb Praxis.* Bern, Frankfurt au Main: Peter Lang.

—— 1995. *Speaking for the Chief: Okyeame and the Politics of Akan Oratory.* Bloomington and Indianapolis: Indiana University Press.

Chapter 10

Indigenous Peoples' Amazonian Sustainable Development Project

Priscilla Settee

I am a First Nations Cree woman, whose community of origin is in remote northern Saskatchewan, Canada. I am the first generation to graduate university and teach there. I have used that position of privilege to open doors for other Indigenous people, notably students and community people. In the late 1970s my teaching position with Canada's first and only Indigenous university opened doors for me to visit and learn from other Indigenous nations in Latin America. While I was a lecturer for Saskatchewan Indian Federated College (later renamed First Nations University of Canada) I travelled for six months to remote Indigenous communities from Mexico to Ecuador and lived among many tribal groups. I was drawn to Latin America because of my interest in the many tribal groups and I felt a sense of commonality with them. It was an eye-opening experience that has defined who I am and the work I do. I could not learn from books only what I learned in the jungle communities of the many Indigenous nations who stories touched me. These were stories of genocide, displacement from homelands and untold stories of human suffering that would later be part of my experience and discourse within international fora such as the United Nations and globalization events impacting Indigenous peoples (Settee 2011). I am a product of the struggle for Indigenous self-determination started by my elders, decades ago, in the Indigenization of primary, secondary and post-secondary education. As an educator I believe that we as Indigenous peoples possess knowledges that have sustained our communities since time immemorial and that this organic knowledge, which finds similar expressions in other regions of the world can be a gift to humanity. I also believe that these knowledges have been expropriated and not acknowledged or compensated as in the case of scientific biopiracy. My PhD dissertation and subsequent book tells those stories from the Pacific, South African regions as well as Canada. I consider myself a change-maker now in that Indigenous struggle as a faculty member in the Department of Native Studies where a small cohort of faculty members carve out a niche in a large western university. An internationalist at heart, in later years my personal educational curiosity and research expanded from Latin America to other regions of the world including the Pacific, Africa and India. It was a natural fit for me to organize and develop the Indigenous Peoples and Sustainable Development Project (IPSDP), the subject of this chapter, that I felt would benefit Indigenous peoples from both the Amazon jungle and later the

Andean highlands of Peru. I admit to operating from a handicap of having too little time and opportunity, to support the development of the project from the ground up as I did not live in the community where the project existed and my command of Spanish was limited. I was also teaching full time back in my home university. Despite its innovative qualities, I believe one of the greatest weaknesses of the project is that a local community on the ground did not develop it and perhaps despite great intentions, it was thought of, framed and introduced to the recipient community. This in itself is a lesson in development education.

In November 2005, my office at the University of Saskatchewan entered into a one year working relationship with the University of San Marcos (USM). The University of San Marcos in Peru experienced problems attracting and retaining Indigenous Amazonian students and so embarked on a collaborative project to address these issues. Initially called the Project for the Formation of Indigenous Amazonian Professionals and Leaders for the Sustainable Development of the Indigenous Communities of the Amazon Region of Peru, it later became known as Indigenous Peoples and Sustainable Development Project (IPSDP). It proved necessary to see the Peruvian Indigenous situation within the broader context of western imperialism, colonization, and development, and the abrogation of international Indigenous human rights. This chapter puts out a pedagogical framework that reflects an Indigenous worldview, to suggest some future directions for Indigenous education in Peru and beyond, focusing particularly on the plight of women. It documents the gains that Indigenous peoples have made within the international arena, in order to focus on conditions within their homelands. Finally, the chapter examines some of the range and depth of knowledge that is currently in jeopardy because of the deterioration of Indigenous lands and cultures brought on by western development. It aims to contribute to the ongoing struggle for social, economic and political justice that Indigenous peoples are currently engaged in.

These thoughts reflect knowledge gained through several decades of teaching at Canadian universities, as well as some 35 years of activism within the Indigenous world, both locally, in Canada, and globally. A central component of the research is the recognition of the serious plight of Indigenous women. This paper derives from a requirement for a doctoral research scholarship received from the Centre for Economic Studies in Lima Peru.

In my experience, Universities and architects of public policy often ignore, or are unaware of, the knowledge that Indigenous students bring from their home communities. Despite this marginalization Indigenous peoples have achieved a great deal in their difficult struggle for recognition.

On my first visit to Lima, the sight of a young mom clutching her baby with one hand, while using a rag to clean the windshields of passing vehicles at midnight in the streets of Lima, has forever been etched in my mind as an image of the conditions Indigenous women are facing in Peru.

An indelible image from a second trip involved a mom holding her baby with her hand outstretched against the backdrop of high-rise buildings of the Lima skyline. This Indigenous woman, in her dusty traditional clothing, sitting on a traffic meridian surrounded by a rush of traffic, was an anachronistic spectre of extreme poverty within the bustling capitalist Lima economy. Another disturbing scene was that of young girls, some no more than eight years old, in the streets at night seeking out men who would give them money for sex. These images can be found anywhere in the Americas, a reflection of the desperate face of grinding poverty amidst extreme wealth, of how particularly women and children are victimized by poverty, and a reminder of the need to link Indigenous development goals to the academy.

Project Partners

The Indigenous Peoples Program (IPP) of the Extension Division at the University of Saskatchewan serves many functions in Indigenous language retention, knowledge and culture by offering community development courses, leadership training, conference organizing, and publishing. The IPP was the initiator and a founding member of the Canadian Indigenous Languages and Literacy Development Institute (CILLDI), which is committed to developing and promoting University level Indigenous language/knowledge courses. I have served as director of IPP, and as an advisor in many regions of the world, working with Indigenous peoples on human rights, land rights, and sovereignty struggles. My MEd from the University of Manitoba looks at Indigenous contributions to the field of western science and my PhD from University of Saskatchewan examines the contributions of Indigenous knowledge systems of South Africa, the Pacific region, and North American First Nations. With this rich and varied background, IPP was an excellent partner to explore the issue of Indigenous student university retention in Peru.

The University of San Marcos (UNMSM) was founded on 12 May 1551, is an academic community and at the time this article was written it comprised of 31,695 undergraduate, 3,495 graduate students, 3,271 faculty and 979 administrative staff. It is located in Lima, Peru.

Peruvian Indigenous Higher Education

Peru is a multiethnic, multicultural, and multilingual country with a population today of 26 million, of which more than 40 per cent is Indigenous. The Amazonian Indigenous comprise 3.19 per cent of this number, and they are the poorest community in Peru. The Peruvian Amazon covers 74.4 per cent of Peru's territory. It has great diversity, both ecological and cultural, with 65 different ethnic groups in 1,450 native communities'' with a total population of over 320,000. They occupy 59 per cent of the Amazon region. Currently, the Indigenous people living in the Peruvian Amazon lack financial, professional and technical resources. The Peruvian Amazonian Indigenous cultures and regional identity are severely challenged by the sustained development which is occurring (Inter-American Development Bank 1997; World Bank/Barclay 1998; World Bank 2000; Smith 2003).

In Peru only a small number of Amazonian young people have access to, and benefit from a university education, and this impacts their home communities. At the beginning of the project, 114 Amazonian Indigenous students drawn from 14 ethnic groups were studying at USM, but many of them lack the academic skills necessary to survive. The university is a foreign and isolating experience for many Indigenous students. Issues such as a lack of the economic means to provide safe housing, good nutrition, medical coverage, books and school supplies pose problems for the students. These all contribute to the success or failure of the students. Added to these challenges is the lack of a national policy in Peru for Indigenous development or bilingual intercultural education for the Indigenous people, and current education policies pay virtually no attention to the position of the Indigenous woman (World Bank/Barclay 1998). Despite some small but significant efforts, Indigenous education remains largely neglected (World Bank/ Barclay 1998).

Discrimination against Indigenous Life

Indigenous Peruvians live as second-class citizens and experience discrimination and human rights abuses, even within their traditional territories. This is a barrier to human rights and to the democratic development of Peru:

> The daily experience of ethnic discrimination must be understood in the context of the particular characteristics of Peruvian society. It is not simply a question of condemning extreme cases, because ethnic discrimination is contained in a web of imperceptible daily events to which we are quite accustomed but which in reality undermine the development potential of those discriminated against. (Acosta and Ciurlizza 1997: 21)

According to the same report, many sectors, including government, avoid the subject of discrimination against and subordination of Indigenous Peruvians.

> From the most open repression to government attitudes of paternalism, the insertion of indigenous people in the official life of Peru has never been complete. There is the problem of open exclusion, but there is also one of attitudes derived from the deep social and ethnic conflict wrought by the Spanish Conquest in Peru. (Acosta and Ciurlizza 1997: 28)

Varese, for instance, refers to the powerlessness of Indigenous Peruvians as a direct lack of political and decision-making power.

> Therefore, such factors as the socio-economic situation of the native society within the national framework, and its differential access or lack of access to the means of control of political and decision-making power necessarily enter into my analysis. (Varese 1972: 3)

Galeano associates the situation in Peru with the marginalization of all Latin American countries under the political and economic domination of global government and capitalism, particularly the United States.

> What happened to Latin America's industrial bourgeoisie was what happens to dwarfs: it became decrepit without having grown. Our bourgeois of today are agents and functionaries of prepotent foreign corporations. Truth compels us to admit that they never did anything to deserve a better fare. (Galeano 1997: 208)

Imperialism continues to benefit the developed countries at the cost of the autonomy and sovereignty of Indigenous peoples. If development does not benefit Indigenous peoples then they are viewed as obstacles or they are seen as being in the way of progress. Even aid is problematic for Indigenous peoples which creates more requirements and restrictions than benefits.

> It is ready to give us its "help" in the form of loans that increase the volume of our unpayable external debt. Nevertheless, after some fifty years of the functioning of the global developments enterprise one notes that no one has achieved the hoped for benefits, apart from the West, which has made considerable financial gains. (Grillo 1998: 192)

The marginalization faced by Indigenous Peruvians continues into higher learning. It may be that there is not yet a critical mass of Indigenous peoples identifying with their cultural roots within these institutions. Attitudes to Indigenous languages illustrate discrimination against Indigenous peoples in Peru, as do those regarding their contributions to the national identity. (Garcia 2003) When commenting on the fact that Indigenous parents would rather send their children the Alliance Francaise or to the North American Institute in Cuzco rather than Quechua-language schools.

> If Quechua were privileged the situation might be different, and we might even want our children to read and write, the conditions for the improvement of our children's education are still determined by our reality. (Garcia 2003: 8)

This discrimination is certainly not manifest because Indigenous peoples have not been active 'on the ground' and within community, (for they have been) but perhaps their stories and actions need to influence different spheres, such as universities, as Cunningham notes.

> Because what we do every day is of no use if we don't create institutions that
> provide follow up and sustainability to our achievements. It is not enough to find
> a government official with good intentions if there is no change in the laws, if no
> defined public policies remain in institutions. This is what determines change.
> (Cunningham 2004: 19)

The current marginalized position of Peru's Amazonian Indians, similar to most
American Indians, has historical roots featuring exploitation and domination. For
more than 500 years, Europeans saw this region as one that provided raw material
for European industries (Smith 1982). Indians of the region provided labour for the
rubber industry, never as owners but always as labourers. The domination continues
to the present. 'Nowhere have indigenous people broken free of that domination to
re-establish their own independent state and civilization' (Smith 1982: 20).

Stavenhagen (1999) sees development of the Amazon region is intimately
linked to the Ethnic Question, that is to say, the long struggle of the Amazonian
Indigenous people and their organizations for their autonomous development,
and the control and management of their lands, territories and natural resources
(World Bank/Barclay 1998). Their regional and national organizations have their
own vision of Indigenous development. In 1998 the Amazonian Indigenous people
elaborated their National Strategic Vision to the year 2005. The Arahuac people
of the Central Rainforest have their Regional Strategic Vision drawn up to the
year 2010 (World Bank 1998, 2002). While the project and this paper do not delve
into the details of both documents clearly both are central to goal of Indigenous
development on the ground and within higher learning.

According to the World Bank Country Study: Poverty and Social Development
in Peru, Indigenous (native) peoples are the most vulnerable and marginalized,
due to unequal distribution of services such as higher education and health care.

> The distribution of social and anti-poverty expenditures has been disappointing.
> The distribution of 7.6 billion soles (about 40 per cent of the total public budget
> in 1996) is mildly tilted towards the better off in Peruvian society; i.e., the poorest
> obtain less of these expenditures than their population share. In large part this is
> due to the anti-poor distribution of higher education and hospital expenditures.
> Several specialized Government programs reach only a small proportion of the
> poor and direct public transfers play a significantly smaller role than private
> transfers do. (Poverty and Social Development in Peru 1999: vii)

This discrimination against the poor has worsened the poverty of Indigenous
peoples, and the situation has been deteriorating over time, rather than improving.

> And we now find that even economically the native population has fallen further
> behind: while in 1994 an Indigenous family was 40 per cent more likely to be
> poor than a non-native family, in 1997 they were almost 50 per cent more likely
> to be poor. (Poverty and Social Development in Peru 1999: 2)

Recent national legislation changes to the Constitution ignore the rights of the Indigenous people and the spirit of ILO Convention 169 ratified by Peru (World Bank/Barclay 1998: 40–41). Today Amazonian Indigenous peoples do not even enjoy access to much of their traditional territories because of mineral exploitation and lumber extraction. (World Bank/Barclay 1998: 38) Hindering their access to natural resources makes it difficult for Amazonian Indigenous peoples to continue to practise traditional economies utilizing traditional knowledge.

It is important for Amazonian students to know the history of their region and its relation to international affairs, situated within an ongoing paternalist colonial mindset. It is equally important for them to learn the strengths of their culture and their identity within the wider society. It is when Indigenous people see their personal, political, economic and social situations within the broader context of domination that they become critically aware of the reasons for their marginalization, the impact of political domination and see education as a potential tool for liberation.

One long term goal of the USM project is to help alleviate poverty, through education of not only students but the public sector as well, in order to raise awareness about the Indigenous socio-economic and cultural conditions. A second goal is the recognition of the knowledge base that is an inherent part of Indigenous culture and their basis for survival.

Indigenous Knowledge in Canadian and Global Context

One of the greatest oversights in both higher education and policy formulation, as it relates to Indigenous communities, has been the lack of recognition and thereby the exclusion of Indigenous knowledge. There are many definitions, of Indigenous knowledge, depending on whether one is a scholar or is community-based. As a body of knowledge, IK has gained currency in the last 20 years among researchers and governmental agencies as well as civil society organizations. African scholar, Odora Hoppers (2010: 9) defines IKS as one that includes, technology, philosophy, education, legal, social, economic and governance systems particularly those of liberation struggles. Indigenous peoples view IK as something that has sustained communities since time immemorial. In Canada, Dakota elder Ken Goodwill argues that Indigenous knowledge is valid in its own right and does not need to be validated by other knowledge systems. Collectivity is central to Indigenous being in the Americas, and the collectivity of Indigenous knowledge is reflected in many of the ceremonies and teachings. Aboriginal scholars and local people describe Indigenous knowledge with labels that reflect ancient knowledge for community life, well-being and sharing values. In the Canadian Cree language this is called *pimatisiwin*. A core value is *miyo-wicehtowin* which means having good relations. Individually and collectively people are directed by their IK heritage to strive and conduct themselves in ways that create positive relationships. The Cree refer to the concept of all my relations, which extends to all of humanity and all living things, as they are related and must be cared for by one other. Over the last four decades Indigenous peoples have worked to create international connections which allow us to exchange stories from our varied communities' Indigenous knowledge bases, and to build strategies for the mutual strengthening of our often isolated nations (Settee 2007). The Canadian International Development Agency defines IK as follows: Indigenous knowledge represents the accumulated experience, wisdom and know-how unique to cultures, societies, and/or communities of people, living in an intimate relationship of balance and harmony with their local environments. These cultures have roots that extend into history beyond the advent of colonialism. They stand apart as distinctive bodies of knowledge, which have evolved over many generations within their particular ecosystem, and they define the social and natural relationships with those environments. They are based within their own philosophic and cognitive system, and serve as the basis for community-level decision making in areas pertaining to governance, food security, human and animal health, childhood development and education, natural resource management and other vital socio-economic activities (CIDA document, no date). Some see IK as a last hope in implementing a sustainable future. We know our 'development' literature overlooks IK.

Others have defined Indigenous knowledge as local knowledge that is unique to a culture or society, outside the formal educational system. It is the basis for decision making in health, agriculture, food preparation, natural resource management and education. Communities, rather than individuals, hold this knowledge. It is embedded in community practices, rituals and relationships and is difficult to codify. IK is part of everyday life Standing Conference of Eastern, Central and Southern African Library Associations (Snyman 2002: 101). Few Indigenous nations enjoy much support from their nation states and instead are forced to spend time negotiating with governments to ensure that basic human rights are met.

Currently some research and development agencies require consideration of women's Indigenous knowledge. Women of the Americas are frequently at the forefront of identifying the challenges and solutions to Indigenous community life and nationhood. It is women who keep the home fronts and communities intact, often in desperate circumstances. Within the last decade the global network of 'Indigenous Women of the Americas' has organized four conferences for networking, discussion and planning for the future of their communities. In April 2004 Indigenous women of the Americas met at the *One Continent, One Spirit* conference in Lima Peru to map out a plan of Action. Women from 22 North, Central and South American countries came together for the third such meeting. The director of the UN Population Fund for Latin America and the Caribbean identified issues such as maternal health, the reduction of maternal and infant mortality, universal education, education for girl-children, overall poverty (which is now critical), HIV/AIDS and the environment (Saavedra et al. 2004: 15).

In my experience after decades of working with I feel they have a clear understanding of both the roots of poverty and marginalization, as well as a variety of solutions to the problems:

> From the perspective of Indigenous Peoples, the impact generated by globalization is devastating and deeply negative since it involves violation of territories, degradation of natural resources, forced relocation of communities, desecration of holy places, daily injustice shown towards their claims, discrimination and lack of respect for diversity and for their rights. The economic policies of this world show no human face. (Saavedra et al. 2004: 45)

Peru is known as one of the great centres of agriculture and food production in the world, attributable largely to Indigenous peoples. There is tremendous opportunity to utilize related Indigenous knowledge, especially in the context of higher learning and policy development. Many of the world's foods, such as potatoes, beans, corn and squash, as well as many plants with medicinal uses, originated from this country. This information needs to be recognized and taught as official curriculum to Indigenous Peruvian students, and indeed to all students. Indigenous peoples were the first to develop the 'planting' method (as opposed to the Old World 'broadcasting' method) of planting seeds (Weatherford 1988). Corn was adapted to grow with a protective husk, which saved the corn seeds from both drought and insects. Prior to their move away from dependence upon wheat-based food to potato and other foods from the Americas, many Old World people died in famine, when wheat crops failed due to disease and drought. Potatoes were a gift developed by Peruvian Indians. In the tiny jungle community of Genaro Herrera, South American Indians are teaching scientists how to cultivate a wide variety of yams, potatoes, and tubers. These Western-trained scientists have no knowledge of many of these plants; 'The scientists working at Genaro Herrera strive to unravel the complex technology of native agriculture and food processing as much as they strive to understand more about the biology of the plants themselves'(Weatherford 1988: 82). Similarly, the 'Center of Indigenous Cultures of Peru' in Lima has, for 20 years, been making a concerted effort to revitalize and, in some cases, reintroduce, many Indigenous foods, including heritage potatoes and beans, through their food security projects. Still, Indigenous peoples struggle to have their contributions acknowledged and compensated (Settee 2007: 8).

The Impact of Globalization on Indigenous Knowledge Systems, Community Rights and Community Development

It is necessary to view the situation of Peruvian Indigenous in respect to the larger picture of globalization, in which Indigenous lands have been exploited for their resources, resulting in devastating consequences for the peoples' well-being. Such experiences have mobilized Indigenous peoples to fight for traditional concepts of sustainable development. Minority Rights Group International states that, without the full participation of Indigenous peoples, the goal of implementing strategies for sustainable development cannot be met. The cultures of the Indigenous peoples of the Americas promote the preservation of the environment and sustainable use of natural resources. They sense that their very existence is under threat while securing few benefits if any to lands they have long occupied.

> The negative effect of harmful and unwanted natural resource development on these communities is striking and constitutes a clear violation of their human rights. In some cases, it is now a threat to their very existence. (Lennox 2012: 12)

Recently the weight of international treaties and contractual obligations to global financial institutions, has contributed to the deterioration in the development performance of a large area of the world over several decades. Often the most vulnerable are blamed for the failures of the development policy establishment. It is imperative to help developing communities meet the challenges by designing national policies consistent with their stages of development and capacities to implement them.

Indigenous Peoples' Global Participation and Networks

Since the first United Nations World Summit on Sustainable Development in Rio de Janeiro in 1992, Indigenous peoples have participated in many international sustainable development and benefits sharing events, including the drafting of the Convention on Biological Diversity (CBD), which was established in 1992 to address biodiversity concerns. It is a work in progress and has been signed on by over 100 countries, including Peru and Canada, but not the United States. The drafting process has been characterized by unequal partnerships. The wealthiest G7 countries have access to and are represented by lawyers, paid government officials and large contingents of support staff, which ensures that their interests are well represented. Even though the US is not a signatory it wields considerable power within the process. The Indigenous peoples have few resources to attend and participate fully in the meetings, and have even fewer official delegations to negotiate their rights and concerns within the Convention's framework. Peru participates in the United Nations forum on Indigenous peoples and has ratified ILO Convention 169. Peru is also a signatory of the International Treaty on Plant Genetic Resources for Food and Agriculture. However, this does not translate into specific policies for the conservation, protection and exploitation of Peruvian or other Indigenous peoples' biogenetic resources and traditional knowledge.

Since meeting in November 1997 in Madrid, Spain, Indigenous Peoples have met regularly on a more organized basis to express their concerns with the Convention on Biological Diversity and more particularly Article 8(j). Article 8(j) which states that a government:

subject to its national legislation, respect, preserve and maintain knowledge, innovations and practices of Indigenous and local communities embodying traditional lifestyles relevant for the conservation and sustainable use of biological diversity and promote their wider application with the approval and involvement of the holders of such knowledge, innovations and practices and encourage the equitable sharing of the benefits arising from the utilization of such knowledge, innovations and practices. (Convention on Biological Diversity 2001: 8)

This is typical of much of the wording of the Convention on Biological Diversity (CBD) that puts Indigenous concerns secondary to nation states' interests. Because they are not recognized as parties to the CBD, Indigenous peoples lack full participation rights in relation to the drafting process of the Convention. Indigenous peoples have further concerns that the CBD does not recognize the relationship that exists between Indigenous peoples, their knowledge, their lands, and biodiversity. With the nation states and the interests of transnational companies controlling them, Indigenous peoples have little or no control over their lands and natural resources. Parties to the Convention do not address biopiracy (theft of flora and fauna) or uncontrolled access to genetic resources on Indigenous lands. There is no linkage between article 8(j) and other international instruments, which address the rights of Indigenous peoples (for example the Draft Declaration on the Rights of Indigenous Peoples). Such concerns led to the establishment of the open-ended working group on Article 8(j) and related provisions of the CBD, with the mandate to establish guidelines and mechanisms for their application and implementation. Its other mandate includes promoting international cooperation between Indigenous peoples and national governments and international organizations. The working group also addresses the implications of matters of prior consent, fair and equitable sharing of benefits and *in situ* conservation on Indigenous lands, meaning the protection of species within Indigenous communities.

It is proposed to protect Indigenous knowledge using the concept of intellectual property rights. But the idea that intellect is a commodity, which can be owned and protected through legal means, is unacceptable to Indigenous communities. The Trade Related Intellectual Property Rights (TRIPS) Agreement of the WTO has caused much debate among Indigenous peoples around the world. On the one hand, Indigenous peoples believe that they must be present where the protection of their knowledge through non-Indigenous methods is being discussed. But on the other hand, the premise of protecting their knowledge through property rights contradicts community rights. While recognizing that their knowledge is being expropriated by outside interests, Indigenous peoples ask how can something collective like knowledge, and what is inextricably at the heart of what makes community, be owned privately. For the most part Indigenous peoples have chosen to participate in order to have their views understood because to not do so will inevitably result in undermining their sovereignty and rights to their natural resources. From a personal Cree perspective I have present as an invited member of the Canadian CBD delegation and have faced frustration with other members of the official delegation who clearly make a point of recognizing Canada's interests before the interests of Canadian Indigenous peoples.

At the international level Indigenous peoples have been mobilizing to educate themselves and others about the international development process and associated financial agencies such as the World Bank, the World Trade Organization and the various international agreements which they believe are at the root of underdevelopment on their lands. Although more recently the pace of work has picked up, Indigenous peoples have been involved in researching, organizing and voicing their concerns about conditions in their homelands since before the first Summit on Sustainable Development in Rio de Janeiro in 1992, (Settee 1996). Indigenous peoples have worked on the Declaration on the Rights of Indigenous Peoples, which represents the voice of over 400 million Indigenous peoples globally, for over two decades. Although it is not law, policy makers must give full consideration to the Declaration, as it represents the accumulated human rights effort of many Indigenous peoples globally. At the United Nations preparatory commissions in Nairobi and Bali, Indigenous representatives produced papers called 'Indigenous Peoples' Caucus Statements for the Multi-Stakeholder Dialogue on Governance, Partnerships and Capacity-Building' and 'Indigenous Peoples' Caucus Opening Statement on Capacity-building'. Both statements address the global social and ecological crises, power relationships between the state and Indigenous peoples and the problem of unsustainable development. Other issues include the need for capacity building and self-governance for Indigenous communities and the need to build on prior work.

> The contemporary world is characterized by deep imbalances in our social relations, of gross inequalities between nations and within societies, manifested by huge disparities in consumption of natural resources; international governance gives disproportionate power to the same economic elite and their institutions of choice – the World Trade Organization and the international economic and financial institutions – to decide the futures of our children. (Indigenous Peoples' Caucus Opening Statement on Capacity-Building 2002: 12)

The statement calls for respect of Indigenous peoples' territories and self-determination, sustainable development at all levels, the need to address self-development and corporate accountability, and cultural and intellectual property rights in regards to Indigenous traditional knowledge.

In September 2003, at the parallel forum to the Ministerial Meeting of the World Trade Organization in Mexico, the concerted efforts of the least developed countries put the process of world trade negotiations on hold. In an amazing display of dogged determination they stopped the developed countries from what LDC referred to as an undemocratic process until further negotiations could address their concerns. Many civil society organizations were present to witness and support the event. At the later Cancun meeting Indigenous representatives from a variety of organizations and NGOs developed a statement that outlined their concerns, which included many of the issues discussed above that Indigenous peoples have been raising at international meetings for some time.

Which Way Forward

It is apparent from the various activities and declarations that have been produced that a critical mass of Indigenous peoples have a clear idea of the lack of democratic processes that greatly impact their communities and they are prepared to challenge them. One way that we as academics can lend support and take leadership is to address some of these issues in our classrooms and in our research and writing. The USM is one way that the north (Canada) was able to lend support through a government funded university directed project.

The USM project was three-pronged: firstly, to assist students to become better students by raising their awareness; secondly, to raise awareness through public and university workshops intended for professors, as well as students; thirdly, to make students aware of the wealth of Indigenous knowledge that they bring from their communities.

The project sought to achieve these objectives by strengthening the 'Indigenous Association of University Students of the Peruvian Amazon' through a Students' Centre for Amazonian Indigenous students. This centre would be a defender of their student rights and a locus to promote their reflection, discussion and exchange on key themes linked to Indigenous peoples, the region, and the Amazon Basin, as well as hemispheric Aboriginal themes. This helped enrich and update Amazonian Indigenous students on the reality of Indigenous people of the Amazon and the entire region.

Challenges and Lessons Learned

On a technical level it became apparent that the logical framework elaborated for the project had an approach to the problematic of the Amazon that was inadequate and not adapted to the reality of the students. The levels of specific objectives, results and activities did not show a coherent approach with the educational cultural and socio economic reality of the Amazonian indigenous students. The majority of the resources assigned did not match with the goals of the project and were reassigned. This situation could have been prevented had the host community/communities been involved with the development of the project from the beginning. The goals were adapted to meet the needs of the students. Workshops, seminars and focus groups were developed to deal with leadership, attitudinal change and sustainable development of the native communities.

While they were invited, it was difficult to get the Consultative Committee made up of AAUPI, AIDESEP, PRATEC and CILA to respond to requests for meeting. Again they were added after the project proposal was developed and the project between both universities had begun.

Despite these major and acknowledged oversights students were assisted in very foundational and material ways, such as food, clothing and housing. In addition, academic assistances such as tutorials for students were very successful. Tutors would work as guides of the students, to ease their transition to the educational system and to orient them in their studies. Finally, without any formal evaluation and with my limited ability to be present at the project site a number of lessons were learned that would serve as lessons to similar projects. Despite the fact that the Canadian government provides funding for similar projects, based on this experience, time needs to be spent with the Indigenous communities to learn what it is they want for their children in terms of higher learning. A similar process needs to be undertaken with the host university within the targeted country. It was clear from the beginning that the organizations that were invited were not interested as it was not their project and they spent little time supporting it. This was true for both the community based Indigenous organizations as well as the administrative bodies and faculty members of USM. Despite the fact that I was one of the co-directors and that I am Indigenous does not necessarily translate into making a project work in another largely Indigenous nation such as Peru. Some of the methods and materials used in the development of Canadian Indigenous education could not simply be transplanted into another Indigenous community in the world given the limited time and language barriers. In many Canadian communities where educational transformation has occurred it is because Indigenous community leaders took the initiative and followed it through despite the often discouraging and unwelcoming messages from university spaces. Indigenous leaders had a desire for their youth to benefit from higher learning. It continues to advance because the Indigenous academic pool is growing and developing some dynamic initiatives often against all odds such as shrinking budgets and growing pressures of globalization. The larger areas of curriculum change, action research and Indigenization of the academy were not even broached in the USM project. To use the analogy it was like fitting round pegs into square holes. I believe that this was largely due to the fact that while the onsite, Peruvian co-director was highly educated, she was neither Indigenous nor was she an academic member of the USM. She did not have the background to lead a major pedagogical initiative although she made some good recommendations based on her observations for actions that would support the students. Some time was needed to develop a cohort of university based faculty that could work as allies in making the academy more user friendly for Indigenous students and to suggest ways that the USM could respond to Indigenous needs. Many workshops were organized intending to educate 'the native' but rarely were the Indigenous stories workshops attended by university faculty. Given the marginalization of Peru's Indigenous and low rates of university attendance, it is apparent that this university is not serving Indigenous. However, without a formal evaluation of we will never know how many lives were improved and how many fell by the wayside. One thing is certain and given the ecological and political climate of Peru's multi-Indigenous communities, reflective of many other places in the Indigenous world, an informal brand of learning, teaching and political activism is rising. Whether the formal institutions such as universities are ready remains to be seen.

The project began to tell the Indian story and it is apparent that so much more needs to be learned from the 'Indigenous' of Peru. We need to hear from more stories of hope and rebirth as Cook-Lyn states:

> Does the Indian story as it is told now end in rebirth of Native nations as it did in the past? Does it help in the development of worthy ideas, prophecies for a future in which we continue as tribal people who maintain the legacies of the past and a sense of optimism? (Cook-Lynn 1998: 134)

Note

After phase one the Indigenous Peoples Program was the recipient of second phase funding for the project Indigenous Peoples and Sustainable Development. The second phase of the project expanded to include the Andean region.

References

Apfell-Marglin, F. and PRATEC (The Andean Project for Peasant Technology) (eds) 1998. *The Spirit of Regeneration: Andean Culture Confronting Western Notions of Development*. New York: Zed Publishers.

Acosta, G. and Ciurlizza, J. 1997. *Democracy in Peru: A Human Rights Perspective.* Lima, Peru: International Centre for Human Rights and Democratic Development.

Convention on Biological Diversity 2001. *Secretariat of the Convention on Biological Diversity, Handbook of the Convention on Biological Diversity*. London: Sterling (VA); Earthscan Pubs.

Cook-Lynn, E. 1998. American Indian Intellectualism and the New Indian Story, in *Natives and Academics: Researching and Writing about American Indians*, edited by D.A. Mihesuah. Lincoln, Nebraska: University of Nebraska Press, 111–38.

Freire, P. 1970. *Pedagogy of the Oppressed*. New York: Herder and Herder.

Galeano, E. 1997. *Open Veins of Latin America, Five Centuries of the Pillage of a Continent*. New York: Monthly Review Press.

Garcia, M.E. 2003. The Politics of Community: Education, Indigenous Rights, and Ethnic Mobilization in Peru. *Latin American Perspectives*, 30(1), 70–95.

Grillo, E.F. 1998. Development or Decolonization in the Andes? *The Spirit of Regeneration: Andean Culture Confronting Western Notions of Development*, edited by F. Apfell-Marglin and PRATEC. New York: Zed Publishers, 124–45.

Indigenous Peoples' Caucus 2002. Opening Statement on Capacity Building of the Indigenous Peoples' Caucus. A declaration by the Indigenous Peoples' Caucus.

Inter-American Development Bank 1997. *Latin America After a Decade of Reforms*. Washington: Inter-American Development Bank Flagship Publication, Economic and Social Progress Report (IPES). Available at: http://idbdocs.iadb.org/w. Retrieved January 14, 2009

Odora Hoppers, C.A. 2002. *Indigenous Knowledge and the Integration of Knowledge System: Towards a Philosphy of Articulations.* Johannesburg, South Africa: New Africa Book Pty.

Saavedra, E.S., Colens, M.V. and Laird, A. 2004 *Memory: One Continent, One Spirit*, IV Continental Meeting of Indigenous Women of the Americas, Chirapaq, Lima, Peru.

Settee, P. 1996. Honouring Indigenous Science as a Means of Ensuring Scientific Responsibility. MEd thesis, University of Manitoba.

—— 2007. Pimatisiwin: Indigenous Knowledge Systems, Our Time Has Come. PhD dissertation, University of Saskatchewan.

—— 2011. Travels Through Global Indigeneity: Testimony of a First Nations Scholar, in *A Heart of Wisdom, Life Writing as Empathetic Inquiry*, edited by C. Chambers, E. Hasebe-Ludt, A. Sinner and C. Leggo. New York: Peter Lang Publishing, 165–72.

Smith, R.C. 1982. *The Dialectics of Domination in Peru: Native Communities and the Myth of the Vast Amazonian Emptiness*. Cambridge, MA: Cultural Survival Institute.

Snyman, R. (ed.) 2002. *From Africa to the World – The Globalization of Indigenous Knowledge Systems*. Pretoria: Proceedings of the 15th Standing Conference of Eastern, Central and Southern African Library Associations.

Stavenhagen, R. 1999. *Structural Racism and Trends in the Global Economy*. The International Council on Human Rights Policy Review Meeting: Racism: Trends and Patterns in Discrimination Geneva, 3–4 December 1999. ICHRP commissioned this document as a working paper.

Stavig, W. 1999. *The World of Tupac Amaru, Conflict Community, and Identity in Colonial Peru.* Lincoln and London: University of Nebraska Press.

Stokes, S.C. 1995. *Cultures in Conflict, Social Movements and the State in Peru.* Berkeley: University of California Press.

Varese, S. 1972. *The Forest Indians in the Present Political Situation of Peru.* Copenhagen, Denmark: International Work Group for Indigenous Affairs. Available at: http://www.chirapaq.org.pe/htm/segurset.htm. Retrieved March 14, 2009

Weatherford, J. 1988. *Indian Givers: How Native American Transformed the World.* New York: Crown Pubs.

World Bank 1999. *Poverty and Social Developments in Peru, 1994–1997.* Washington, DC: The International Bank for Reconstruction and Development.

—— 2000. World Development Report 2000/2001: Attacking Poverty. Oxford: Oxford University Press.

—— 2002. World Development Report 2002: Building Institutions for Markets. Oxford: Oxford University Press.

Chapter 11

Engagement and Ownership of Knowledge: Issues Affecting Indigenous Education and Pedagogy

Raymond Nichol

Indigenous peoples' criticism and distrust of some anthropological research, and anthropologists' struggles to remain relevant and effective in decolonized settings, mean a theoretically and methodologically ethical anthropology is imperative. In Australia and Melanesia, community members often complain of cultural knowledge and artworks being exploited, of promises from researchers not met, of careers made while the 'informants' remain impoverished. Research needs to be more responsive to the community and involve members fully and collaboratively in the process, from beginning to end. Does this mean we get too close to the people and the issues affecting them, often so disastrously? Can 'engaged anthropology' be objective and analytical? As Paul Sillitoe asks the reader, does such local empowerment threaten scholarly integrity, objectivity and academic authority? I argue in this chapter that such empowerment, rather than threatening, is essential for best practice in community education and development. I make many recommendations, drawn from extensive ethnographic experience. Teachers' self-fulfilling prophecies of their Indigenous pupils and the low self-esteem of so many Indigenous learners demand change in current practice. I've heard teachers state to 'town' children, 'You can do better than that, you're not from the mission', while girls in the playground derided each other with, 'You nothin' but a black crow'.

Of particular relevance to the discussion in this chapter, focusing on Indigenous education, is the way schooling in Australia and Melanesia is caught up in the system of stratification at local and national levels. Reflecting a decolonizing research agenda it draws from ethnographic, historical and comparative Indigenous education research studies. The field is so complex, and often fraught, that there is a need to take account of shifting interdependencies of factors influencing Indigenous learning and the need to recognize the tensions in any decisions made about the relative importance of how these component factors are represented. These are taken into account when untangling complex pedagogical approaches and systems.

Factors of race, class, status, economic and political power, affect all who participate in schooling. While education systems in Australia and Melanesia (for this chapter most examples come from Papua New Guinea), facilitate some individual social mobility, making potential leaders and others more confident in dealing with outsiders, they also place people in particular racial, class and status categories. They organize and structure the life-chances of both their beneficiaries and victims and legitimize their place in systems of unequal privileges and rewards. Hence, schools, despite the best efforts of many principals and teachers, do not usually challenge the local system of social stratification. Ultimately they function, in the main, to preserve and legitimate the existing inequalities of power, wealth, status and land ownership.

Incorporating Indigenous Knowledge and Pedagogy into Education

The chapter responds to contemporary policy and practice in Indigenous education. It recommends to all educators and community developers, in Australia, Melanesia and elsewhere, that they incorporate elements of Indigenous knowledge and pedagogy into their organization of learning, classroom practice and project development. For the ethnographic case studies see Nichol (2006) and, for more detailed educational explanation, Nichol (2011). Culture and education are inextricably interwoven since the content of all education has value underpinnings that are always associated with a particular cultural agenda (Thaman 2001: 1).

> Indigenous knowledge is a growing field of inquiry, both nationally and internationally, particularly for those interested in educational innovation. The question, "What is Indigenous knowledge?" is usually asked by Eurocentric scholars seeking to understand a cognitive system that is alien to them. The greatest challenge in answering this question to find a respectful way to compare Eurocentric and Indigenous ways of knowing and include both into contemporary modern education. Finding a satisfactory answer to this question is the necessary first step in remedying the failure of the existing First Nations [Canadian] educational system and in bringing about a blended educational context that respects and builds on both Indigenous and Eurocentric knowledge systems. (Batiste 2002: 3)

Underlying this chapter are crucial contemporary issues of reconciliation between Indigenous and other Australians and the lack of a sense of national identity and citizenship of people from the various provinces and social strata in Papua New Guinea. It suggests some strategies for educators and community developers dealing with these social dilemmas. Elements of reconciliation and citizenship range from the wording of preambles, constitutions and treaties, to social, economic, legal, cultural and political justice and acceptance. There are numerous related issues and dilemmas across this vast region. Appropriate education is vital. People learn better together if they know and appreciate something of the others' pedagogical background. To feel comfortable and confident when learning is crucial to outcomes in education. The chapter draws on extensive ethnographic and teaching experience in schools, universities and Indigenous communities, in Australia and Papua New Guinea, a wide review of literature in the field, plus critiques and suggestions offered by eight Indigenous teachers from the Yipirinya School, Alice Springs, Northern Territory, Australia, one of whom commented that:

> In my experience, most mainstream schools don't cater for a diverse range of students, preferring to teach in a mainly white, middle class fashion. Students who come from a different culture or background are expected to assimilate, or else face a difficult learning situation, which could lead to them eventually "dropping out" of school.

Consequences of the Imposition of Western Education

After the invasion and occupation of their lands by Europeans, Indigenous peoples were expected to benefit from a Western education. The benefits were seen to be apparent, particularly in comparison to the perceived 'stone-age cultures', 'heathen' beliefs and customs, and in Australia the 'squalor and horrors of the black camp'. Thousands of Aboriginal children, particularly those of mixed race, were taken from their parents, to be assimilated to European ways. The inadequacies, often horrors and abuses, visited upon many of those children constitute another story (see the *'Bringing Them Home' Report*, Wilson 1997). Many of those who remained with their parents were provided with a second-rate education that, somewhat incongruously, tried to integrate them while denigrating their language, culture and community.

This is not an exclusively Australian or Melanesian experience. Educators in other regions, from all levels of the education system, need to compare the way minorities in their country have fared in education and society. For example, in Japan the *Burakumin* 'untouchables', the Koreans, Indigenous *Ainu*, and even *Kikokushijo*, Japanese returnees from extended periods overseas. All experience an educational system that is, at times, less than supportive of their needs. Studies of Indian, African, Native American, South American, Maori and Sami societies, also provide telling examples of Indigenous systems of learning and knowledge being ignored or derided. The immigrant experience is often similar.

A major concern in education has been the lack of relevance of much of the content and methodology imposed upon Indigenous students. Gradually some educators have realized that Indigenous children learn differently and that their culture and pedagogy have validity and strength. Of course, educators also need to be acutely aware of the diversity of Indigenous cultures, particularly in Australia and PNG, and that there is not a monolithic sense of identity or pedagogy. Dispossession of land, alienation, poor health and few employment opportunities must also affect ownership of knowledge, educational interest, attendance, application and performance. Issues of rights to make decisions for the community, particularly about the land and community development, also affect educational outcomes. The most appropriate and effective learning strategies are explained as being holistic, imaginal [or imaginable], kinaesthetic, cooperative, contextual and person-oriented (Craven 1996, 1999; Nichol 2004, 2006, 2011). To ignore key social and environmental aspects of learning, as too often occurs, is seen as being particularly damaging for marginalized Indigenous and minority students at all levels, in all places.

Relevance and Application

Many teachers say they struggle when trying to teach Western concepts and methods to Indigenous students. This raises many relevant issues, such as communication, initial literacy, language, and ownership, recognition and accommodation of Indigenous pedagogy. Also, if you wish to perpetuate inequalities then provide the *same* educational methodology and content for all. Teachers who tell me that a 'fair go' means treating all the students exactly the same need to know that this is not productive. Of course, if educational provision has been second-rate for Indigenous students, then moves towards 'equality' with mainstream provision are improvements but, too often, are far from engaging or culturally relevant for Indigenous and other minority culture learners. This point was made strongly by another of the Yipirinya teachers when she said, 'I would go so far as to say, to expect one style of teaching to work for a diverse range of students is unequal, unjust and could be deemed as racist'.

Contemporary Indigenous culture, in Melanesia and Australia, is complex and diverse, from traditionally oriented people living in isolated communities with little contact with the outside world, to people living and functioning ably in predominantly urban, post-industrial contexts. For example, at Yipirinya School, based in suburban Alice Springs, there is considerable diversity among the students. There are more than four different local language groups and the students come from 'bush communities', 'town camps', and from Alice Springs. Aboriginal English is a *lingua franca*. A Yipirinya teacher explained the students' backgrounds as being:

> "Town Campers" – these are students who live in small Aboriginal communities in and around Alice Springs. They can still speak their languages, but are increasingly influenced by the western ways. Most of these students face an overwhelming swag of difficult issues, such as living with alcoholism, petrol sniffing, domestic violence, child abuse, racism, poor health and poor housing.
>
> "Bush mob" – students live a more cultural and traditional way of life. Their language and culture is very strong and they speak little English.
>
> "Townies" – these students live in urban Alice Springs. They either suffer from the same issues as the "town campers", or at the other extreme they don't suffer at all. They tend to lack culture and tradition, speak little or no Aboriginal language, but do speak Aboriginal English.

Many of the children living on the fringes of towns and cities in PNG, the Solomon Islands and Fiji, face similar realities of everyday life. One can see why education, if available, suffers. In Australia it is available almost everywhere, and there are many inducements to attend, from primary to tertiary, however standards overall remain low.

Primary schools in Melanesia cater for considerable numbers of young children, however as drastic 'cut-offs' occur in secondary and tertiary education many from remote villages and marginalized settlements are increasingly denied opportunities, for education and employment. Decisions made are not just logistic and economic. As applies to Australia, many students are disengaged and disillusioned by the schooling offered.

Subjects that, by convention, are taught to European standards and by mainstream methods, including Vocational Education Training (VET) offerings, should take into account their relevance to the Indigenous student. Many Indigenous leaders in Australia express distaste for much of what their people are taught. For example, it is galling when children are told that Captain James Cook discovered Australia, and that Australia was settled rather than invaded. Content, as well as pedagogy, needs to be accurate, appropriate and relevant. In Papua New Guinea, particularly in colonial times, readers and textbooks were criticized for not being culturally relevant. In the 1930s the government anthropologist founded an English newspaper, so locals would have something relevant to read in the language of instruction! It has long been bemoaned that English is taught in the school but then, for many, has little further reinforcement or relevance.

Much of what is found in this chapter has wider application in contemporary educational theory, policy and practice, especially engagement. For example, many students of African and Middle Eastern origins experience difficulties with the subjects and methodology offered in mainstream schools. Students whose language and culture is not based on Anglo-American Western customs and heritage often experience difficulties with English literacy, history and the social sciences. Therefore, many of the insights and methodologies proposed to develop more relevant curricula and pedagogy for Indigenous students could be used to make subjects more interesting, relevant and successful for *all* students.

History of Indigenous Education in Melanesia and Australia

In traditional societies young people learned as they grew up, with an informal learning system based, in the main, on need to know, and supported by a more formal system of initiation and organized instruction. This instruction was organized by 'clever', 'powerful' or 'big men', educational and political leaders, whom the anthropologist, A.P. Elkin, termed, in Australia, 'Men of High Degree', (while he acknowledged that older women could also qualify and practise). In central and western New South Wales these teachers were known as *wireenan* and *walamira*. Across Australia and Melanesia they were elders, repositories and controllers of Indigenous knowledge, usually with strong genealogical ties to the learners.

All young people were 'put through the rules', 'broken', 'tamed' or 'steered' through life, and while much learning was observational and incidental, no society left learning to chance. Sanctions for going against 'the law' were serious, from shaming, physical punishment, banishment, to death. Education was organic, heterogeneous, and ensured a complementarity of gender roles (if frequently with fearful, antagonistic, oppositional, violent elements, particularly in highland Melanesia and the desert regions in Australia).

Learning took place, in the main, during day-to-day activities. Indigenous people were often fluent or could 'hear' in a number of neighbouring dialects, allowing communication with surrounding groups. Skills were learned by observation, imitation, real life practice, and from the oral tradition, linking song (stories, legends, instruction), site (place, land, property, fishing, hunting, gathering rights), skin (kinship, family, lineage, obligations) and ceremony (rituals, dancing, instruction and ties to the past). This led to the following characteristics of traditional Indigenous education.

In brief, and recognizing significant variation across the Australasian region, learning was largely oral and the use of storytelling was important. Sign language was frequently used. Education was largely informal, except during initiation, when formal, even coercive, and rigorous methods of instruction were employed. Initiates later referred to being 'ritually killed and born again', 'tied in', 'broken into' or 'steered' through initiation. Less formal methods included observation, imitation and casual instruction. Learning occurred through participation in the life of the community. Often instruction came as people gathered around a fire, leading to the Aboriginal phrase, 'We grow them up in the ashes'. Everywhere in Aboriginal Australia, the hearth, the family or community fire, constituted a place for gathering at the end of the day, where food was shared, stories told, songs sung. Seating arrangements around the fire were significant in terms of the location of each person's land (sitting in the direction of 'country') and those with whom he or she could be close or distant, generous or practise avoidance. Through these means, a rich cultural heritage was transmitted and children learned the social, economic and religious life of the community, including philosophy, ethics, art, music, dance and mythology. Religion (perhaps better, spirituality) permeated every aspect of life. There was no purely secular education.

> [In Australia] All hunting, food-gathering, family life and social life were intimately connected with their religious belief. (Hart 1970)
>
> [In Melanesia] Orokaiva religious beliefs and many of their medical practices centre on two spirit concepts, *asisi* and *sovai*. ... According to the Orokaiva, all living things, animals and plants, have *asisi*. The *assisi* of things may impinge on human life in many ways [for example, causality]. The *sovai,* in contrast [are spirits of the dead]. (Sillitoe 1998: 218–28)

Education was closely adapted to the mostly subsistence economy. It prepared learners to hunt, fish, farm, build houses and boats, tools and artefacts, gather and track, gain knowledge of the seasons for fish, animals, fruits, tubers, sago, the location of water holes, methods of obtaining water from certain tree roots and plants, and so on. It was life-related and life-inspired. Children learned social responsibilities associated with relationships and the significance of certain individuals: often for boys they were father's brothers (often referred to as 'father'), and for girls, mother's sisters (often referred to as 'mother'), and other close relatives. Knowledge and experience of the kinship system was central to learning. Sometimes the learner initiated the process. For example, a person wishing to learn a particular craft would observe a specialist over quite a long period. When ready in his or her mind the 'apprentice' would manufacture the artefact, usually to a high level of replication and quality. Education extended throughout life. Stages of wisdom were acknowledged according to age, and status in the community.

After European Occupation

After the European occupation of Australia and Melanesia, Indigenous students were gradually introduced to European, Dutch, German or British/Australian, formal education. Efforts were directed to convert the Indigenous 'heathen', 'savage', 'stone-age' culture and to replace traditional learning with 'superior', dominant European knowledge.

Initially Indigenous children, and adults to a lesser extent, were taught a basic form of Western education that allowed them to function as servants and workers for the European colonists. Advanced studies were thought to be '… beyond the natives' grasp'. Until the 1930s they were considered, on an official, governmental level over the whole region to be virtually ineducable. As late as the 1960s Charles Barnes, Australian Minister for Territories, thought that Papua New Guineans 'might be ready for self-government in a hundred years'. Paul Hasluck, a previous minister, had been more positive concerning the country's potential for independence, so there was no unilateral position in Australian government. However, Barnes did state that Australia would implement a timetable for an independent Papua New Guinea if a party with a coherent program for self-government were successful in the 1972 House of Assembly elections. To the surprize of many, one of whom was undoubtedly the Minister, Michael Somare and his Pangu Party duly won the election and formed an effective coalition. A popular book around the time of Barnes' comment was *Kiki: Ten Thousand Years in a Lifetime* (Kiki 1968), which incidentally contains some brave and barbed criticism of colonial attitudes persisting into the 1960s. These attitudes persisted in the wider community until recently, and remain in pockets of Australian and expatriate PNG society.

The curriculum until the 1960s was, in general, one deemed appropriate for the lower orders and, by all the evidence now available, usually failed miserably. The mid to late 1960s and 1970s in PNG signalled an enormous, last ditch push to prepare an elite for self-government and independence. It was they, the new ruling class, who were provided with vastly disproportionate levels of resources, time and energy. In the main, they, their children and grandchildren, still are.

Attitudes towards those 'Missing Out'

In Australia and Melanesia, notwithstanding the reaction of some conservative elements – particularly in privileged and expatriate groups in Papua New Guinea and in rural Queensland, Western Australia and the Northern Territory – there is greater appreciation today, in terms of policy at least, by society and government of the worth of *all* citizens.

In Australia, there is a considerable dismantling of societal barriers to advancement. Aboriginal students have greater access to education, with positive community support through Local Aboriginal Education Consultative Groups (LAECGS) and government support. For example, Victoria's Koorie ('Koorie' is a term many Indigenous people in Victoria, Tasmania and New South Wales prefer to 'Aborigines') Open Door Education (KODE), later Koorie Pathways schools. These are expensive and some have problems with attendance and results, however they indicate strong state support for Indigenous educational initiatives.

In Papua New Guinea, national and regional agreements in education have led to some success, particularly wider access to primary and tertiary education. As then Prime Minister Michael Somare stated in 2010,

> Education in PNG is a success story. And many great things have happened in our country but we have been conditioned to focus on the not so good and forget the leaps and bounds we have made. We can compare ourselves to the rest of the world and be discouraged or we can compare ourselves in terms of our own history and be encouraged to do more. It is a simple truth, that there have been more people educated in the last 30 years than ever before in PNG's colonial history. When I led the country to nationhood there was just one newly established university (UPNG) and very few university graduates around me. Today we have six universities, many colleges and other tertiary institutions for our five million people. And I have around me many well-qualified men and women who serve the country in many different ways, both in the private and public sectors.

This is obviously putting the best possible construction on the present situation in the country. As for Australia, there are many dysfunctional Indigenous communities, particularly in fringe settlements, with low levels of gainful employment, even for those with some formal education. In remote communities, there is likely to be no access to western/national schooling. Change is desperately needed, especially to counter the sense of 'otherness', appropriation, powerlessness and lack of inclusiveness and citizenship. Too many teachers in the Australian/Melanesian region, particularly in secondary education, carry a hangover of 'superior' Western or dominant culture, knowledge and methods, a feeling that only they know what is 'correct to know'. They are wedded to the methods that they have used for many years. Their cultural background and training usually ensure this is so. Often they teach as they were taught. The traditional Western classroom, particularly in secondary schools, has a teacher who explains or demonstrates an issue, problem or concept, provides some examples and then sets the students to work on problems or issues of a similar kind. These problems are graded, from simple to difficult. Often they are designed to tie in other learning, concepts and themes. If they give a justification for the topic, issue or problem it is usually, 'It's on the Essential Learnings, the Standards, or the Framework' (the national, state or regional-approved curriculum framework). The content is largely derivative and positivist. Standardized testing ensures conformity and 'teachers teaching to the test'. Teachers, textbooks and standard computer software are, in reality, seen as being the font of wisdom.

In Australia, past history and social studies courses, indeed even in the relatively recent Curriculum Standards Framework documents in Victoria, there is little sense of 'blackfellas' blood on the wattle and billabong', of savage 'clearing the pastoral run'. The destruction, dispossession, segregation, attempted assimilation of Indigenous Australians, even land rights, educational and leadership successes, are often ignored. There *are* some admirable Indigenous Studies courses, particularly at primary and upper secondary levels, but delivery is patchy and methodology often questionable. Community relations, reconciliation and efforts to 'close the gap' are not necessarily enhanced. Clearly content, as well as pedagogy, needs to be accurate and relevant.

> To ignore content is to ignore why Indigenous students choose not to access mainstream schools. The learning is not relevant to them or their lives. They don't see the value of an all-inclusive white education, especially if it is at the expense of language and culture. Content is equally as important as pedagogy and learning styles. (Yipirinya teacher, Alice Springs)

In Papua New Guinea it has been argued from the nineteenth century to the present day that a national education system should be more responsive to both local and national needs, and, in particular, be more appropriate to the economic circumstances of the country. 'In a few short years the school system seemed to have moved out of step with the occupational structure and was creating potentially dangerous social problems' (Smith (ed.) 1987: 262). Thousands of Standard 6 and higher school graduates find themselves '... not useful either in the village or in outside employment ... they seek refuge in the larger towns of this country ... [where] if they are lucky they find work. But most remain unemployed for long periods of time … these young men start on another kind of secondary education – that of cowboy films, pubs and suffering' (Smith (ed.) 1987: 263). Today, one could add 'rascal' crime, drugs, violence, and so on.

The clans of these young people are crucial to their futures as the social units where customary education occurs. The 'Melanesian Way' of decision-making, allocation of resources and resolution of disputation, shapes, in part, the political culture that pervades the state. Therefore, customary ways of doing things must be linked more closely to schooling and pedagogy.

Links to Contemporary Constructivist Learning

Constructivist, inquiry-based approaches encourage the student to engage with and discover the law or concept by experiment with tactile, relevant and contextual teaching aids. They untangle and integrate, linking well with traditional approaches to learning, and tending to be more responsive to the interests and needs of the student. They are more group-oriented, conceptually creative, integrated, holistic, and conducive to solving problems. Teachers have less dominance of the learning process. In Indigenous settings, the teachers' planning involves liaison with Aboriginal students, staff, parents and the Local Aboriginal Education Consultative Group. Ideally, students and staff work together cooperatively, using field research and community experiences, computers and the Internet, as well as book research, to allow real-life data with cultural and pragmatic relevance, to be processed and evaluated more easily and effectively.

The ethnographic and school-based research indicates that these forms of pedagogy and teaching encourage Indigenous and other students to have ownership of their learning and to take far more interest in their subjects, general learning and even school attendance. For example, the Indigenous students in Gippsland, Victoria, went from having the highest truancy rates in Gippsland to the lowest when they began attending their community controlled and supported school. The principal, chosen by the community, was non-Indigenous, but the Indigenous cultural and personnel 'presence' in the school was strong. Parents and other community members were frequently in the school, supporting the students and teachers. This school was impressive in its use of information technology, frequently exploring and presenting Indigenous themes.

The Indigenous Child as Learner

If we are to develop and implement an Indigenous pedagogy for we require an appropriate framework for learning, based on sound anthropological and educational research, responding to the needs of Indigenous students and communities. Of course, I am cognizant of the dangers of over-generalization, reductionism, dichotomous thinking, and 'tips for teachers' (see Nicholls,

Crowley, and Watt 1998). I do not intend the following pedagogical model to be prescriptive for all Indigenous students, as a model for separating Indigenous from other Australians or Melanesians.

The following characteristics of Indigenous pedagogy, grounded in ethnographic case study, come predominantly from the research and writings of Craven ((ed.) 1996, 1999), Nichol (2004, 2006, 2008, 2011) and Main, Fennell, and Nichol (2000). They are also influenced by complementary ways to develop cross-cultural dialogue and education, Indigenous and Western, namely the 'two way' and *Ga<u>n</u>ma* 'both ways' systems (see Harris 1990; and Creighton 2003, respectively). In sum, 'best practice' learning is holistic, imaginal, kinaesthetic, cooperative, contextual and person-oriented.

> Today, the … way we think and learn largely depends on our ability to clarify for ourselves the differences between our received wisdom (from our formal, mostly western education) and the wisdom of the (home) cultures in which we grew up and were socialised, and from which we continue to learn important knowledge, skills and values. (Thaman 2001: 1)

Holistic is interpreted as meaning conceptually integrated and all-encompassing. Indigenous children tend to prefer holistic or integrated approaches to learning. It reflects traditional worldviews in which everything is interrelated and all relationships are important. It also reflects the importance of family and place. When elders are asked how a sense of identity, of Aboriginality, or of being a *wantok*, a villager, Busama, Manus, Wogeo, and so on, is acquired, they often say something akin to, 'We grow them up in the ashes. That is, our children learn around a campfire or hearth, in the bosom of their family and kin (or the men's or women's house)'.

> At Yipirinya school awareness of relationships is acute among the teachers and students. This allows students to feel safe and happy and therefore able to learn. (Yipirinya teacher, Alice Springs)

Holistic learning approaches do not compartmentalize learning according to academic disciplines. Areas of study are concurrent and integrated so that learning flows smoothly between content areas, and the interrelationship between knowledge and skills is apparent. Students prefer to observe and discuss a task or topic before working through the components and activities. They learn more effectively if the overall concept and direction of a lesson is outlined, discussed and modelled before engaging in activities.

> The children tend to learn better when they can make the connection and relate it to the whole concept, as opposed to looking at concepts in an isolated manner. It has a more real life approach and is more reflective of their Indigenous worldview. (Yipirinya teacher, Alice Springs)

This is particularly significant for the early years of learning; however secondary and tertiary teachers should also endeavour to integrate learning more and to apply concepts and analysis across disciplines.

Imaginal learning is relatively unstructured and consists of thoughts, images and experiences of learning. As for holistic learning it is strongly linked with notions of identity, perhaps expression of Aboriginality, or in Papua New Guinea, of being a *wantok* or citizen. In Indigenous contexts, learning occurs more frequently in informal, unstructured ways, through observation and imitation rather than verbalization.

Cultures without a literate tradition are strongly auditory, as shown by their strong oral traditions, however, interestingly, there is little verbal interaction for the deliberate and conscious purpose of teaching and learning. There is often a tradition of oratory at ceremonial gatherings, which may well have an educative function as it frequently has a berating element! However, information is transmitted primarily through observation and involvement.

In imaginal contexts, images are also a more effective means of communicating knowledge and keeping students engaged. Imaginal learners may have difficulties with purely cognitive operations. They learn more effectively if concrete examples precede abstract understandings, employing visual images, symbols, diagrams, maps and pathways. Their uncanny skills in football and other positional sports may derive from this way of being in the world.

> Aboriginal students form pictures of tasks in their minds and then perform them through imitation. They prefer to see the "whole" rather than "little bit by little bit". In this way they have the task and the expected outcome and are then prepared to give it a go ... They often need concrete materials to conceptualize what they need to learn. For example, when teaching a social studies lesson we might take students on a "bush tucker" excursion. (Yipirinya teacher, Alice Springs)

Exclusively teacher-centred instruction (that is, 'chalk and talk') is not an effective form of instruction for imaginal learners. However, there is a place for teacher-centred instruction at times, particularly if the class has a common misconception or misunderstanding.

Kinaesthetic learning is tactile, through manipulation and movement within the learning environment. Many Indigenous students are kinaesthetic learners. Information is taken in more easily through their hands and through movement. They like to handle things, to move them around, to also move around themselves. As noted above, they are often talented 'play-makers' in games and sports, anticipating and moving into an ideal position, seemingly effortlessly.

One of the most effective social and environmental education strategies for kinaesthetic learners is to develop excursions and tasks where students, in working groups, collect data outside the classroom; '... culture trips and country visits provide an excellent opportunity for this to occur' (Yipirinya teacher, Alice Springs). The data is recorded in notes and photographs for later application at school or home. This is a key factor in the success of a number of school programs, being enjoyable, engaging and often challenging. The participants can later share their ideas and projects with family and other students. Models, including computer models, websites, dioramas, sculptures, tableaus, and artistic project presentations, tap this propensity. Computer games, simulations, computer skill, dexterity and attractive presentation activities, also build on this attribute.

Cooperative learning places emphasis on communal, shared and group learning. As the ethnographic research with the Wiradjuri and Wongaibuwan of central-western New South Wales reveals (see Nichol 2006), Indigenous cultures, for reasons based on their very survival as hunter-gatherers, often place a higher priority on the group than the individual. Learning usually takes place in groups and is collaborative. Cooperation is more important than competition or individual achievement. Therefore, in classrooms, students who have group discussions, interpretation of instructions and assistance, are more likely to be successful. However, there is a place for individual and more formal teacher-led tuition.

> Currently I am employed at Yipirinya School as a literacy and numeracy tutor for Grade 1 and 2 students. In this role I deliver tutoring on a one-to-one basis. This is quite the opposite to what Nichol advocates in terms of "best practice" teaching strategies. Whilst I agree that group work is preferable and Aboriginal students are happier in this situation, at times one on one is necessary to strengthen their skills. I develop a very strong, personal connection with each child. In this way they feel comfortable and supported. Similarly "our room" is within the main classroom so they still have a sense of being part of the group. (Yipirinya teacher, Alice Springs)

Pointing at and singling out a child in the main classroom, even for praise, may be seen as confrontational. Those who respond with downcast eyes, and what may appear to be a sullen expression, are not necessarily showing you disrespect. By contrast, in non-Indigenous society teachers usually deliver instructions with strong emphasizes on competition, individual benefit and achievement. Looking the teacher in the eye and answering confidently, directly and openly, is praised and rewarded. Some students, including some Indigenous students, thrive on this; many do not. As Australian Council of Educational Research (ACER) research reveals,

> Indigenous students and non-Indigenous students differ in their learning styles. Indigenous students are more likely to be cooperative learners, whereas non-Indigenous students are more likely to be competitive learners. This finding would suggest that appropriate and effective pedagogical and assessment practices for Indigenous students would be ones that [incorporate] Indigenous students' learning style. (Mellor and Corrigan (eds) 2004: 35).

Specificity and relevance, that is placing content and pedagogy in context, are crucial to effective learning; 'I find this to be of particular importance at Yipirinya ... many students come from bush communities and town camps where much of their time is spent outdoors. It is essential that this be translated into the teaching situation. We hold many classes outdoors and out bush. In the language classes, students are allowed to move about freely ...' (Yipirinya teacher, Alice Springs).

In traditional societies, learning occurs in the specific context to which the learning relates. Children learn hunting techniques during food gathering on land and sea, songs and dances during community celebrations, kinship responsibilities by interaction with relatives, artefact manufacture by long observation with minimal verbal instruction. By contrast, Western schools are usually more artificial learning environments, where content is removed from and often has little apparent application, to daily life. By placing information, activities and learning in context, students discover that education is meaningful and relevant to their own lives. The 'expanded horizons' approach used in social education programs often has contextual value. Knowledge and skills acquired through local studies are later applied in wider contexts. For example, studies of chemicals in the home may lead to students researching industrial applications of chemicals. A study of a local vineyard or sheep/wheat farm might be contrasted with swidden agriculture in Papua New Guinea or 'factory' farming in China.

Indigenous cultures are more person-oriented than information-oriented. By developing person-oriented learning, we emphasize that family and personal relationships are the key to positive learning outcomes. Teachers are assessed by how they relate to the children and community rather than by their qualifications or performance as instructors. My observations, and own teaching experience, indicate that students who feel a personal connection with the teacher will be more cooperative, interested in learning, willing to take risks and attempt new tasks. 'This specifically relates to a feeling of "family-ness". At Yipirinya teachers take a particular interest in each student; they get to know their families and become part of their lives' (Yipirinya teacher, Alice Springs).

Fieldwork experience made me very aware that peppering my discourse with 'please' and 'thank you', and other English gentilities, was often disconcerting for Indigenous students. In many communities favours are expected because of on-going reciprocity. It is neither expected nor necessary to offer thanks. It is likely that, over time, the student will acquire some of these cultural niceties, as many a visitor has in other cultures.

Students will work well with, and for you, if you establish positive relationships with them and clear understandings of reciprocity. Tangible reinforcement is better than verbal. If teachers are rigid about excessive politeness and formality then they risk a breakdown of communication with their students. I have seen and heard of many teachers with a 'shape-up or ship-out' approach, causing intense resentment and alienation among students and their families.

It is often advisable to accept higher levels of 'working noise' in the classroom and use non-verbal strategies to regain attention. Also, it is enjoyable and valuable to work out a sign-language system understood by all. Organize the classroom furniture with quiet areas and areas for group activities to give students more control of their own learning. Teachers can improve student achievement through simple strategies such as acting positively and consistently, welcoming students warmly to class, and building self-esteem through positive reinforcement and learning places with Indigenous symbols and references. Many teachers and Indigenous leaders say that Indigenous children are highly skilled readers of body language. In a Northern Territory Aboriginal community education survey most students defined a good teacher as, essentially, 'Someone who likes us and is fair'. Many had experienced dislike and discrimination, in their own country.

> As long as … society contains (at least) two ethnic cultural traditions, one identifying with those who were colonised and one with those who did the colonising, then there will inevitably be conflicting attitudes … [If we] … are to share a common citizenship this needs to be anchored in some shared values, such as justice, and some shared traditions, such as egalitarianism and the "fair go". (Pearson and Sanders (eds) 1998: 193)

Learning Outcomes and Assessment

Assessment can be very confronting, particularly for Indigenous students, so the teacher should aim to use methods with which the students are comfortable. Include assessment tasks that allow students to demonstrate their knowledge visually and physically rather than just in verbal and written forms. Try fostering expression of curricula concepts and themes by using environmental and immersion language techniques, drawing on the students' own experiences.

Assess comprehension by having students retell the activity, task or story using movement and facial expression. Use assessment that rewards teamwork. Avoid alienating students through criticism, particularly in the early years, or with new concepts or skills, by trailing the introduction of self and group-assessment of work.

> As an Aboriginal person myself, going through school and now further study, I would say there is no comfortable way at all. Assessment is confronting and the Aboriginal student will either stay away or choose not to even attempt it. I would suggest an on-going form of "hidden" assessment, inbuilt in all lessons as a way to get around this. I would suggest that a teacher may have "failed" in her teaching if a student has "failed" in assessment. (Yipirinya teacher, Alice Springs)

Of course, if student attendance is poor, as it often is, then one can hardly blame the teacher, particularly if the school is welcoming and well run.

As the students will face more formal assessment during upper secondary and tertiary study I suggest gradually preparing students by introducing small class tests, 'open book' at the beginning. Short dictation tests allow for immediate feedback and assistance, with the bonus of checking whether students are 'hearing' effectively. Also, many computer-based quizzes and tests prepare students well for more formal testing.

> At Yipirinya School we have programs that cater for … students' needs, such as: nutrition, personal hygiene, health worker visits, and specialist health such as hearing, eyesight, dentist. We also offer four separate language classes, country visits and cultural trips for each language class. (Yipirinya teacher, Alice Springs)

While cultural sensitivity and knowledge is essential for non-Indigenous staff, there is much about the wider societies and their political and economic systems that Indigenous people need to know if they are to be empowered and confident enough to embrace citizenship. They need an introduction to the 'dominant culture', so they can deal effectively in the wider society. 'Two way' and *Ganma*, 'both ways', mean just that. Using the 'expanded horizons' approach studies might lead from the family, community organizations and local governance, to state, federal and international politics. At all levels provide local, relevant experiences and information *as starting points.*

It is advisable to use multimedia resources, including Internet, computers, and so on, to explore and demonstrate concepts. The websites of Victorian schools are particularly illustrative of how creative and talented Indigenous students are when provided with such opportunities (http://www.sofweb.vic.edu.au/Koorie/index. htm). For Papua New Guinea the following site is helpful: http://en.wikipedia.org/wiki/List_of_schools_in_Papua_New_Guinea#Schools.

Students need to be able to use their everyday literacies to learn the new literacies of contemporary schooling, verbal, visual, graphical and numerical. They need to connect learning to their everyday worlds and values.

Conclusion

> Indigenous pedagogy should be embraced by all teachers, and, indeed, all students would benefit from this. In terms of reconciliation this is only one part, but it is certainly an essential one. (Yipirinya teacher, Alice Springs, Northern Territory)

My research, teaching and experience in this field, indicate that Indigenous students and their families must not continue through the new century seeing schooling as being alien, appropriating and threatening, as they have too often in the past. If the strategies presented above are trialled and implemented, then educators, Indigenous and non-Indigenous, will be far more likely to assist their

Indigenous students to negotiate their place in their nations, the global economy, and a world of technological change.

We can all learn a great deal from the Indigenous world. While we should be aware of the diversity within Indigenous societies, the strategies recommended have wide application. In fact all teachers and others working with Indigenous communities can learn much from Indigenous pedagogy. As Paul Sillitoe states, Indigenous knowledge is 'a unique formulation of knowledge coming from a range of sources rooted in local cultures, a dynamic and ever changing pastiche of past 'tradition' and present invention with a view to the future'. Development cannot be meaningful unless Indigenous knowledge is integrated into the development process (2002: 113).

Those who have 'grown up in the ashes' should have their cultural background acknowledged and catered for, and experience success in education and in their lives as fully participating citizens of their own country. This is the hope and demand of Indigenous and minority communities everywhere.

References

Batiste, M. 2002. *Indigenous Knowledge and Pedagogy in First Nations Education: A Literature Review With Recommendations*, National Working Group on Education and Ministry of Indian Affairs, INAC (Indian and Northern Affairs Canada), Ottowa: Ontario, Canada.

Beresford, Q. and Partington, G. (eds) 2003. *Reform and Resistance in Australian Education: The Australian Experience.* Nedlands, WA: University of Western Australia Press.

Craven, R. (ed.) 1996, 1999. *Teaching the Teachers: Indigenous Australian Studies for Primary Teacher Education.* Sydney: University of New South Wales Press.

Craven, R. and Marsh, H. 2003. Teaching Pre-service Teachers Aboriginal Studies: What Really Works? AARE Conference, New Zealand: 20 November–4 December.

Creighton, S. 2003. The Yolngu Way: An Ethnographic Account of Recent Transformations in Indigenous Education at Yirrkala, Northeastern Arnhem Land. Canberra, ANU, PhD thesis.

Harris, S. 1990. *Two-way Aboriginal Schooling: Education and Cultural Survival.* Canberra: Aboriginal Studies Press.

Harrison, N. 2008. *Teaching and Learning in Indigenous Education.* South Melbourne: Oxford University Press.

Hart, A.M. 1970. A History of the Education of Fullblood Aborigines in South Australia. University of Adelaide, MA thesis.

Kiki, A.M. 1968. *Kiki: Ten Thousand Years in a Lifetime.* Melbourne: Cheshire.

McRae, D. et al. 2000. *Education and Training for Indigenous Australians: What Has Worked (and Will Work Again)*. Deakin, ACT: Australian Curriculum Studies Association.

Main, D., Fennell, R. and Nichol, R. 2000. Reconciling Indigenous Pedagogy and Health Sciences to Promote Indigenous Health. *Australian and New Zealand Journal of Public Health*, 24(2), 211–15.

Nichol, R. 2004. 'To Grow Up In The Ashes', Responses of Indigenous Teachers to a Pedagogy for Social Education. *The Social Educator*, 22(1), 6–18.

—— 2006. *Socialization, Land and Citizenship Among Aboriginal Australians: Reconciling Indigenous and Western Forms of Education*. New York: The Edwin Mellen Press.

—— 2008. Learning From Indigenous Worlds: A Case Study and Comparative Analysis of Indigenous Pedagogy and Education. PhD thesis, Faculty of Education, La Trobe University.

—— 2011. *Growing Up Indigenous: Developing Effective Pedagogy for Education and Development*. Rotterdam, Boston and Taipei: Sense Publishers.

Nichol, R. and Robinson, J. 2000. Pedagogical Challenges in Making Mathematics Relevant for Indigenous Australians. *International Journal of Mathematical Research in Science and Technology*, 31(4), 495–504.

Nicholls, C., Crowley, V. and Watt, R. 1998. Theorising Aboriginal Education. *Education Australia*, 33, 6–9.

Pearson, N. and Sanders, W. 1998. *Indigenous Peoples and Reshaping Australian Institutions: Two Perspectives*. Canberra: Discussion Paper No. 102, CAEPR (Centre for Aboriginal Economic Policy Research).

Purdie, N. et al. 2000. *Positive Self-Identity for Indigenous Students and its Relationship to School Outcomes*. Canberra: Department of Education, Science and Training.

Schwab, R. and Sutherland, D. 2001. *Building Indigenous Learning Communities*. Canberra, ANU: Centre for Aboriginal Economic Policy Research, No. 225.

Sillitoe, P. 1998. *An Introduction to the Anthropology of Melanesia: Culture and Tradition*. Cambridge: Cambridge University Press.

Sillitoe, P., Bicker, A. and Pottier, J. 2002. *Participating in Development: Approaches to Indigenous Knowledge*. London: ASA Monographs, Routledge.

Smith, P. 1987. *Education and Colonial Control in Papua New Guinea*. Melbourne: Longman Cheshire.

Somare, M. 2010. *Prime Ministerial Statement on Challenges and Successes in Education for Papua New Guinea*. Port Moresby.

Thamen, K.H. 2001. Towards Culturally Inclusive Teacher Education with Specific Reference to Oceania. *International Education Journal*, 2(5), 1–8.

—— 2003. Decolonizing Pacific Studies: Indigenous Perspectives, Knowledge and Wisdom in Higher Education. *The Contemporary Pacific*, 15(1), 1–17.

Wilson, R. 1997. *Bringing Them Home. Report of the National Inquiry into the Separation of Aboriginal and Torres Strait Islander Children from their Families*. Sydney: Commonwealth of Australia.

Chapter 12

Questions of Power in Schooling for Indigenous Papuans

Rachel Shah

Provision for the education of the children of highlanders living in Papua and West Papua[1] is patchy and diverse. State schools have been built in many areas, and are often attended by indigenous Papuans, though their provision is insufficient to meet the national standard of all citizens attending the first nine years of school. Some indigenous communities continue to provide an education for their children without recourse to schools. Others have access to schools that are set up or supported by foreign agencies. This chapter reflects on some of the challenges these schools face in providing formal education for highlander children, arguing for reflection on the criticisms of education and research made by indigenous peoples in other parts of the world, to avoid the same mistakes being made in Papua. I apply an analysis of power issues based on Bishop's (1995, 1998, 2005) model of initiation, benefits, representation, legitimation and accountability to these schools, and argue for the development of power-sharing partnerships appropriate to the Papuan highlands.

History of Schooling in Papua

Until the arrival of missionaries and Dutch colonisers in Papua, learning outside of schools was considered relevant, appropriate and sufficient for the reproduction of knowledge, skills and beliefs in highlander children; the learning was effective in ensuring that children grew into functioning adults within their environment and community. No known distinction was made by highlanders between 'enculturation', a process of learning which may occur unintentionally or subconsciously, and 'education', which can be defined as the deliberate effort to promote learning (see for example Varenne 2007; Nichol 2011). Both took place without learning institutions such as schools.

Over the last century, however, non-formal learning has come to be seen by many highlanders as insufficient preparation for adulthood. The reasons for this are

---

1 Since 2003, the Indonesian western half of the island of New Guinea has been divided into two provinces, called Papua and West Papua. For readability I, like Munro (2013), use 'Papua' to refer to both provinces unless otherwise stated.

many and complex, but are nearly all rooted in interactions with non-indigenous peoples and their institutions. Dutch colonisers, missionaries, anthropologists and explorers are among those who have impacted Papuan highlanders. Indonesian citizens from more heavily populated provinces moved to Papua under both Dutch and later, Indonesian transmigration programmes. The provinces have abundant natural resources which have brought employees of international companies to the region. In more recent years NGOs and government aid programmes have brought yet more foreigners in the name of development. Each arrival has presented new opportunities and threats, and meeting them required skills and knowledge that Papuans did not yet have. The newcomers' education did not equip them to understand and negotiate with the highlanders either. But it was the Papuans who were portrayed as 'uneducated'. In some cases, this was due to ignorance, ethnocentricism and even racism. Butt's (1998: 47–8) story of an elderly Javanese woman describing the Papuan people as 'very primitive' is sadly common. Many people who encounter Papuan highlanders interpret the conditions of their existence as evidence of people who are incompetent, unimaginative and backwards (see Munro 2013). Other reasons for the portrayal of Papuans as 'uneducated' range from a deliberate attempt to dominate them to gain access to resources, to a desire to help them access 'development' such as health care and various technologies.

Some Papuans themselves, whether because of religious conversion, desire for consumer goods, or curiosity, wanted access to the foreigners' knowledge and skills. As Nichols (2011: 78) observes, in some contexts 'medicine, steel axes, knives, guns, tobacco, alcohol, abundant food, aeroplanes and radio (that is, cargo) were such powerful forces that the village elders and the tenants of traditional society were no longer seen as wise and powerful'. Some Papuans realised, too, that if they didn't access the skills that would enable them to negotiate the complex cross-cultural encounters that were rapidly changing the shape of their political, economic and social environments, and do so quickly, they would continue to be positioned as helpless and 'backwards' and thereby lose the opportunity for self-determination and even survival. For a variety of reasons, then, the process of learning already in place for Papuan highlanders was considered insufficient and Papuan children were portrayed as in need of 'an education'.

Consequently, schools were introduced to educate the children of Papuan highlanders in the foreign knowledge. They had differing forms, pedagogies and curricula to meet different initiators' agendas. For example, in the 1950s, the Dutch used schooling to steer a group of elite Papuans towards independence, attempting to prepare them to self-govern (Timmer 2005; Mollet 2007). Meanwhile missionaries set up Christian schools, using a model based on Dutch schooling but which took into account the Papuan context by adapting the curriculum and teaching methods to suit local children (Mollet 2007). Indeed, even where no schools were set up, missionaries frequently taught literacy so that Christian converts could read the Bible. When Indonesia started to govern Papua, more schools were set up, this time based on the western Indonesian system of education, with six years of *Sekolah Dasar* (*SD*) primary schooling, three years of *Sekolah Menengah Pertama* (*SMP*) junior secondary schooling and three years of *Sekolah Menengah Atas* (*SMA*) senior secondary schooling. In some urban areas there was also access to tertiary education. Teaching was in Indonesian and most teachers were Indonesians from other provinces. But education in Papua was not a priority under President Suharto's government and state schools were mostly inaccessible to people in rural areas (Mollet 2007).

In 2001, when Papua received the Special Autonomy Law (*Otonomi Khusus Papua*), responsibility for formal education at all levels was handed over to the Papuan provincial government, with the national government stipulating only the general policy, core curriculum, and quality standard guidance (Special Autonomy 2001). According to the Special Autonomy Law, at least 30 per cent of the provinces' allocation of profits from natural oil and gas should go towards education, designated as one of four priority areas of development by the Papuan provincial government (UNDP 2005). This new investment has made sufficient resources available for many school buildings to be built in rural areas and for many indigenous Papuans to be trained as teachers.

The Current Situation

Regardless of Papua's educational ideals and commitment to finance them, I have been told and frequently observed that many schools in rural areas are not currently running according to stated aims. I have visited four remote areas in Papua since 2011, a Korowai speaking village in the lowlands, a Moi-speaking settlement in the highlands, a large Lani-speaking village called Bokondini in the highlands, and a series of homesteads in a Walak-speaking area of the highlands. Neither the Korowai nor the Moi had a school within walking distance. Bokondini has three state schools, two *SD* and one *SMP*. When I visited Bokondini in 2011, only one of the *SD* was in operation although national exams were imminent and according to local accounts the other two schools had not been open for several months. The Walak area also has two state *SD* and a state *SMP*, all of which are open, though run by *guru honor* (roughly, 'honorary teachers'), people who teach in the place of the permanent teachers (*guru tetap*) who are away from their posts. The *SMP* is well attended, and open every day, but runs for only a few hours a day. Activities include collecting firewood for the teachers, cutting the grass on school land with machetes, and playing football on the large field. *SD* activities are similar, though more sporadic and less well attended. A single *SD* teacher often has a dozen students of all ages in one 'class'. Even when schools are in operation, problems such as ineffective and status-based (often dictatorial) teaching methods and low self-confidence among pupils due to verbal discouragement and physical punishment are common. Reports of teacher-assisted cheating on examinations are also common, as the quality of education does not adequately prepare students for their exams, but test results and the resulting certificates are considered all important. The cost of taking exams at the Walak *SMP* is currently one pig, one chicken and Rp. 250,000 (approximately £16.50) per student. Both expatriate and Indonesian educators that I have met are unanimous that the current quality of formal education is shockingly low in Papua.

Research done for international organisations echoes these observations, reporting serious concerns about school accessibility, high dropout rates, and the lowest levels of student enrolment in Indonesia (Government of Papua Province et al. 2009; The World Bank 2009; UNDP 2005). Illiteracy in 2012 is reported by Indonesia's Central Statistics Body, *Badan Pusat Statistic* (2013) as 34.49 per cent of the population for Papua province and 6.04 per cent for West Papua province. The World Bank (2009) reports 62 per cent illiteracy in Jayawijaya, one of the more urban regencies of the mostly rural highlands. One of the most common problems cited is teacher absenteeism leading to closed schools; teachers often live in urban centres instead of the remote locations of their posts. Reasons for this include poor infrastructure, lack of support and resources for teachers in rural posts, and lack of monitoring (Government of Papua Province et al. 2009; The World Bank 2009; USAID 2009; UNDP 2005).

A prevalent charge over the last four decades is that schools in remote regions do *not* 'conserve the traditional wisdom and genuine identity of the Papuan people' as one draft for a Provincial Special Regulation Concerning Education states that they should (Draft Perdasi 2002: Article 35, Clause 7). The Papuan administration has set out to reverse this over the last 10 years, but in reality has made little, if any, adaption to the national curriculum and pedagogy, which are inappropriate and irrelevant to local Papuan contexts (Government of Papua Province et al. 2009; USAID 2009; Mollet 2007; UNDP 2005; see also Spicer 1972; Ajamiseba 1987; Duijnstee 1972). The curriculum does not incorporate knowledge or content related either to life in the highlands or to job opportunities in the cash economy (Mollet 2007) and uses methods and examples that are beyond Papuan children's frame of reference (UNDP 2005). More importantly, the relevance of the curriculum has little actual bearing on children's experience of schooling in rural areas as it is rarely, if ever, referred to. A child's school certificate usually has more to do with their relatives' ability and willingness to meet teachers' demands for payment and gifts than it has to do with either the curriculum or the child's knowledge and abilities.

Mission and development agencies are stepping in to provide alternatives to state schooling, with so many schools closed or inaccessible to children, and those that are operating leaving children and young adults no better equipped to deal with the challenges, opportunities and threats of their rapidly changing environments than they were before the introduction of state schools. If *quality* schooling can be provided, the argument goes, the next generation of Papuans will be better equipped to deal with these challenges. They will be better able to manage the resources of their region, to interact on a global scale and to advocate for themselves. They will be equipped to negotiate contracts with international companies, to make demands for the rights of their families, and to self-govern effectively within the current social, economic and political context. As Nichol (2011: 34) points out, 'it could be argued that if self-management and self-determination are to be realistic goals for Indigenous communities then the leaders will need to be able to communicate, liaise and negotiate with outsiders and these skills require academic education'. Indigenous peoples in other parts of the world have insisted on their right to education, in which they usually include schooling, as part of their right to self-determination (see, for example, The Coolangatta Statement on Indigenous Rights in Education 1999; Smith 2005; Manuelito 2005; May and Aikman 2003).

There are as many approaches to achieve quality schooling as there are schools trying to realise it. Whilst a number of foreign agencies are trying to support change within the state system, others are investing in private schools. One school, for example, focuses on maths and physics and takes Papuan children to Jakarta (over six hours' flight away) to take advantage of state-of-the-art, purpose-built facilities (see Surya Institute 2009; Somba 2009). Another, funded by the United Nations Development Programme (UNDP), aims to prepare children for school by teaching maths and language skills to pre-schoolers using spears, bows and other familiar tools (UNDP 2012; Sulthani 2012). A third, in the Papuan town of Sentani, hosts children from the highlands in dormitories, aiming to 'equip Papuan children to be national and international leaders' (Sekolah Papua Harapan 2012). A Dutch foundation (see Oikonomos Foundation 2012) works with Indonesian NGO *Yayasan Kristen Wamena* (Wamena Christian Foundation) who run a primary school in Wamena, a town in the highlands, offer teacher training and are developing curriculum appropriate to the Papuan highlands. Yet another school in the highlands, *Ob Anggen,* is run by a cross-cultural team that seeks to actively involve parents in the education of their children whilst providing schooling which meets international quality standards (USAID 2011). All of these schools are well-attended by Papuan children and report encouraging results, although none have existed long enough to measure the long-term effects of their interventions. Even in these alternatives to state education, however, the introduction of schooling as a means to a relevant education for Papuan highlanders' children is rife with complications, which come to light through reflection on the evaluations made by indigenous peoples of their experiences of mainstream education elsewhere, and by the criticisms which have been made about another arena of education that has impacted indigenous peoples' lives – that of research (Bishop 2003).

Analysing Power Relationships in Research and Education

Although the highlands of Papua have not been exempt from the long history of research into indigenous peoples' lives, for a variety of complicated reasons there are fewer Papuan voices challenging the power relations within such research than there are in many other Pacific and Melanesian contexts. Nonetheless, the work of indigenous scholars from other parts of the world is relevant to the Papuan situation. Smith (1999), Bishop (1998, 1999), and Sandri (this volume) pose questions about the locus of power within research about indigenous peoples, and with others pose similar questions about the power relationships at play in the provision of schooling for children of indigenous peoples in Aotearoa/New Zealand (Bishop and Glynn 1999; Porsanger 1994; Bishop 2003).

One helpful model for analysing power relations in education is outlined by Bishop (1995, 1998, 2005; see also Bishop and Glynn 1999: 55) who sets out questions about power in five areas: initiation, benefits, representation, legitimation and accountability. Initiation refers to who begins a project and defines the outcomes of it. For example, many researchers have considered the ethics of their research, such as selection of research topic and field site, access to knowledge, and informed consent, from the perspective of the researchers' institutions only, and some researchers have entered communities without following local protocols for guests which would give the hosts the power to permit or deny access (Harvey 2003; Te Awakotuku 1991; Nakata 1998). Benefits refer to who stands to gain, and at whose cost, by the project. The way that indigenous peoples have been objectified and insulted through some traditional research, which was of no benefit and in many cases does considerable harm, has been widely condemned (for example, Heckler this volume; Porsanger 1994; Te Awakotuku 1991; Smith 1999; Bishop 1998, 1999; Nichol 2011). Representation refers not just to how people in a project are represented but also to who represents them, on what basis and on whose authority. Traditional research has presumed to represent people on the basis of only brief encounters (Smith 1999), from a standpoint which has made contestation difficult for indigenous peoples (Nakata 1998). Maori students have even found their own histories being taught to them according to the representations of foreign researchers who became known as 'experts' about Maori people (Te Awakotuku 1991; Bishop 1998). Legitimation refers to how the knowledge in a project is validated, which is usually according to Western epistemologies and methodologies, including a value for neutrality and objectivity among the positivists, and explicit subjectivity among the postmodernists, both of which depend on a view of distance between the knower and the known (Bishop 1998; Heshusius 1994). Accountability refers to who retains the right and power to determine what acceptable outcomes of the project are, and to who controls the procedures, evaluations and distributions of knowledge (Bishop 2005). Accountability has traditionally been to researchers' institutions, funders and academic peers, rather than to the people who are hosts (Smith 1999; Porsanger 1994; Bishop 1999). Within what Bishop (1999) calls traditional research, power in these five areas is located primarily, and in many cases exclusively, with the researchers and their institutions.

Serious recognition that research is 'indissolubly related to power and control' (Porsanger 1994: 108) has led to new approaches being developed. Within anthropology, attempts to respond to criticisms of ethnographic research have led in some cases to foreign researchers seeking to utilise 'empowering' models of research in which the researcher 'allows' local people to participate in the research and 'gives voice' to them in the final publication. Bishop's (1999) discussion of emancipatory models of research in light of Kaupapa Maori Research, a framework for research that has been developed by Maori people as an alternative to traditional research, draws attention to the ways that 'empowering' research, though well-intentioned, still assumes that the power and control within the research relationship belong to the researcher, who is therefore in a position to 'give' power to members of the researched community. Indigenous peoples have disputed this premise, arguing that power over research should be retained by the people it is about (see also Sandri this volume). Some people have even suggested that only indigenous researchers should be allowed to conduct research about their own peoples. A more widely held view, however, is that there is space for non-indigenous researchers who are willing to participate on indigenous terms (Porsanger 1994; Smith 1999; Bishop 1999; George 1990, cited in Cram 1992). A variety of models for such engagement by outside researchers have been proposed, mostly within the context of Maori research. These include a model in which a non-Maori researcher is guided by an authoritative Maori mentor, where the researcher is adopted into a *whanau* (broadly, a concept which refers to the extended family and contains Maori values and social practices, see Bishop, 1998: 217) and takes on the associated obligations. It is a power-sharing model in which the researcher and community collaborate on the research to meet outcomes that are desired by the Maori community (George, 1990, cited in Cram 1992).

Bishop's (1995, 1998, 2005) model for analysing power relations in education, used so far mostly in the context of schooling for Maoris (for example Bishop and Glynn 1999; Bishop 2003) or research about Maoris (Bishop 1998, 1999), can be applied to schools in Papua, to illuminate the complications inherent in the introduction of schooling for Papuan highlander children.

The Issue of Initiation

Firstly, there is the issue of initiation, including the choice of appropriate pedagogies and epistemologies for use in school (Bishop 1995, 1998, 2005). Some schools for Papuan children have been started by foreign agencies for the benefit of Papuan people; others have been or are being started at the request of particular communities. In both cases, my experience has been that local Papuans are eager to have access to formal education for their children and there is intense competition over limited places. However, even where schools have been set up at the request of a community, there are wider issues of initiation to consider, such as what form a school will take, what pedagogies will be used, who will run it, and what goals it will aim to achieve. Traditional school models, whether borrowed from Australia, Britain, America, or the western parts of Indonesia, were designed to meet needs and serve purposes that may be inappropriate in the Papuan highlands.

Despite the dominance of positivist epistemology in much of the history of Western schooling and research, the existence of different epistemologies is well documented (see Calabrò this volume and Ferraro this volume for explorations of the ways Western social science knowledge is situated culturally and historically). The field of ethnomathematics, which is particularly illustrative because of Western mathematics' reliance on positivist epistemology, provides many examples of the different ways that people think about and deal with number. These include different limits on how high numbers go, the use of different numbers for different categories of things (such as different numbers being used for animate things and inanimate things, for example Borden 2013), and seemingly different ways of handling abstraction with number. These differences have been used to describe people as primitive or childlike (Ascher and Ascher 1986). One of the reasons that differences in use and understanding of number have prompted such ethnocentric perceptions is that mathematical concepts are often, falsely, thought to be natural, logical and independent of context (Jorgensen and Wagner 2013).

When examined more closely it becomes obvious that the differences in the ways people use number are not differences in ability to think logically or abstractly. It is no less logical, for example, to use different numbers for the living and the dead than it is to ignore the qualities of items being counted. A pertinent example is provided by a study done in the 1930s which used syllogisms to investigate the reasoning abilities of non-literate people in Uzbekistan and found that some people either did not answer as expected by the researchers or refused to answer at all (Ascher and Ascher 1986). For example, when presented with the problem 'All Kpelle men are rice farmers. Mr Smith is not a rice farmer. Is he a Kpelle man?' one response was 'If you know a person, if a question comes up about him you are able to answer. But if you do not know the person, if a question comes up about him, it's hard for you to answer' (Scribner 1977: 490, cited in Ascher and Ascher 1986). Ascher and Ascher (1986: 131) translate this logic as: '*If A then Q, if B then not Q, if not Q then not C, given B and not A, the conclusion is not C*'.

The logic is excellent even though it is different to that of the question (the logic of the question is '*If A then C. Given B is not C, is B equal to A?*'). The response demonstrates not only the responder's ability to think logically and in the abstract, but also the substitution of a rationale which is more consistent with the responder's understanding of the limits of knowledge (Ascher and Ascher 1986). In other words, the question is answered from a different epistemological stance. As Ascher and Ascher comment (1986: 130), 'the Kpelle respondent and his Western questioner have different views on talking about people whom you do not know'. Wola highlanders in Papua New Guinea evidence a similar mindset regarding trust and abstract knowledge (Sillitoe 2010).

These examples provide crucial insight into the way that people from different cultures form, use and reproduce knowledge. The problem is that in schools, especially primary schools, it is common for just one epistemology (often a positivist one) to underpin each subject; for example, only one way of dealing with number is taught as valid or 'right'. It is commendable to teach children the positivist epistemology that underpins much of scientific and technological advancement, but to do so to the exclusion of all other perspectives is a loss to students around the world who can go through school without even knowing about the existence of different epistemologies. For children of Papuan highlanders, who may encounter in school the claim that the epistemological stance of their whole family is wrong, the loss, both educationally and psychologically, is even greater.

Pedagogies also vary between cultures, and are intrinsically value-laden (see Nichol, this volume). For example, in many western Indonesian schools children are expected to learn what their teacher tells them by rote (Frederick and Worden 2011; UNICEF Indonesia 2009). The teacher may punish questioning and expects pupils to afford him or her great respect (Frederick and Worden 2011). This reflects and communicates a hierarchical ordering of society and the importance of paying respect to those in authority. By contrast, many American schools reward questioning, curiosity and students figuring things out for themselves, reflecting a value for independent learning and the development of personal autonomy (Alexander 2001). Another example of a value-laden pedagogical approach is the use of competition to motivate youngsters. The value-laden nature of pedagogies becomes obvious when a pedagogical approach that seems natural in one cultural context is used in a new one. For example, how effective are pedagogies rooted in values of hierarchy likely to be in egalitarian societies?

The Issue of Benefits

Secondly, there is the issue of benefits. The schools for Papuan children in which foreign agencies are involved exist explicitly to benefit Papuan people. The questions of *how* Papuans will benefit, according to whom, and at what cost, are important ones to ask though. These issues are very complex in the Papuan context. Many highlanders want and pursue a formal education for their children, citing a variety of potential benefits. Some people argue for formal education as a means for their children to gain prestige, social status and a salaried, influential or high status position in the cash economy, asserting that they are setting them up for an easier life than that of subsistence gardening. Others assert that those who have had a formal education can use the knowledge they have gained for the benefit of their people, by negotiating the political system to gain influence or funds, by facilitating 'progress', often represented by roads, electricity, and communications, by positioning themselves and their people as powerful actors in a modern society, or in support of political freedom (Munro 2009, 2013). Schooling is also seen as a way to 'overcome alleged backwardness' (Munro 2013: 33). Parents perceive formal education as a powerful tool, something to be taken advantage of.

Nonetheless, people engage in complex analyses of the costs and benefits in deciding whether to send their children to school or not, as they are aware that there may be negative consequences to schooling too. Boarding schools are an example: the separation of children from their kin and their land is painful for both them and their communities, and can be destructive to their sense of identity and belonging (Nichol 2011). In schools that are run in the highlands, costs may be more hidden but they still exist. For example, children who attend school lose time that they would have spent in other activities, which may lead to a loss of opportunity to participate in local events, fulfil cultural obligations, and learn indigenous skills and knowledge. In some school settings, students encounter humiliating portrayals of themselves and their heritage, and learn negative attitudes to their own language and culture.

Even when *SD* schools run in rural settings, children usually have to move to town to finish their schooling, provoking fear that away from the influence, support, and intervention of their family members, they will start drinking or smoking, get pregnant or get married, get into conflict, get in trouble with the authorities, get sick – which is often viewed as preventable if family members are around to intervene spiritually – or even die away from home. Anxiety is heightened by those who leave to finish their schooling and return home with '*stres*' – a word used to describe people who, having been pushed beyond their capacity to cope, have suffered an undiagnosed mental breakdown (see also Munro 2009: 153).

Formal education usually comes with a financial cost, too, and the benefits don't always play out as expected. One reason that the potential loss of indigenous knowledge is not always taken seriously is that schools are expected to equip Papuan children to work in the cash economy, in jobs that require the ability to read, write and use information technology. But in the central highlands, where the main source of employment is public service, according to both my Walak and Lani friends, and Munro's (2009) Dani informants, having the right relatives is at least as important to gaining employment as having the right skills and knowledge. Some young people find themselves at the end of an expensive education without the desire to make a life in the rural homesteads they came from, but, due to a shortage of jobs or the 'wrong' connections, without any means of gaining employment. Others take work that they feel ambivalent about, or disappointed with. Some finish their schooling indebted to those who sponsored their education financially, but unable or unwilling to fulfil the obligations consequently expected of them. According to my Walak friends, most people who complete their schooling never return to live on their land.

Munro (2013) argues that 'inflated possibilities of [state] education may be considered a form of violence because these claims set indigenous men and women up to fail amid unacknowledged conditions that make personal, social and political transformation a highly improbable outcome of schooling in highlands Papua'. I would argue that indigenous men and women are aware, before they invest in private or state education, of the conditions they study in and of the different forms of success schooling results in within those conditions, yet many of them still pursue formal education. However, foreign agencies involved in education should themselves acknowledge that access to schooling does not translate directly into benefits for Papuan highlanders. Papuans' perceptions of the benefits of schooling may also differ from those of the foreign agency. Understanding how complex the issue of benefits is will help schools to foster more just learning environments as well as prepare children to deal with the unjust realities they are likely to encounter beyond school.

The Issues of Representation, Legitimation and Accountability

Thirdly, in Bishop's (1995, 1998, 2005) model there is representation, which relates to several different aspects of schooling including materials, staffing and student enrolment. Whose histories and experiences are represented in the educational materials used in schools? Deeply embedded in educational materials are taken-for-granted depictions of reality and of 'normal' childhood experiences. What gender roles, for instance, are represented in storybooks about the family? Do the men and women live in the same or different houses? What jobs do the people in story-based maths problems have? What aspirations do the phrases used for language learning suggest? SIL International are working to develop culturally appropriate materials for some Papuan contexts (USAID 2009: 141) and *Yayasan Kristen Wamena* continue their work developing curriculum and accompanying education materials for the Papuan highlands that are based on Indonesian national standards (Serasi 2012), but educators still have few materials to draw on which represent the day-to-day experiences of any Papuan highlanders, and even fewer specific ones for each context. It is more difficult still, in most areas, to find comprehensive resources that have been developed in the indigenous language. Representation is important in staffing as well as in materials (Kaomea 2005), as materials can be presented in unintended and counterproductive ways to the aims of those who developed them.

A related issue is the question of how beneficial it is to attempt to capture and teach indigenous knowledge through an institution. As Agrawal (1995) has suggested, incorporating indigenous knowledge into formal education is complex and raises dilemmas about what exactly indigenous knowledge is and whether it should or even can be formalised and separated from everyday life (see also Hermes 2005; Ismail and Cazden 2005). Sarangapani (2003) explores interest in indigenising school curriculums in the context of the Baiga, a small tribe who live in forested regions of central India. She concludes that the *ashram shalas* (formal schools) which present knowledge in literary forms, out of context, and expect children to be competitive and to perform analytical and abstract tasks are pedagogically and epistemologically different from *vidya* (indigenous medicinal knowledge) which must never be written down, is secretive, transmitted orally, non-systematised and integrated with Baiga ways of bringing up children. She argues that the survival of *vidya* will be best achieved by keeping it out of formal education (Sarangapani 2003).

Another important representation aspect of schooling is the issue of the identity of pupils. If there are limited school spaces, some method will be used to allocate them, which will determine who is represented in the school, and who therefore may end up with the skills needed to interact in the global political and economic arena. Related to this is the question of whose aspirations and agendas are being represented through the school. For example, the Dutch schooling system deliberately created a 'Papuan elite' – a group of people who, if the Dutch colonisers' plan had worked, would have represented Papua internationally

(Timmer 2005; Mollet 2007). The question of how to decide who has access to the skills that can be learned in school, and of who should make such decisions, can be difficult even without the expectation that the schooled children should become leaders who represent other Papuans in interactions with the rest of the world. When that expectation is present, the complications of whether or not other Papuans want to be represented arise, including how they want to be represented and who they want to represent them. A particularly fraught question is whether schooled Papuans are still *representative* of their families and communities (see Sillitoe this volume). A common criticism of schooled Papuans is that the process of education has so alienated them from their communities that they can no longer speak for them (a problem indigenous Maori scholars have encountered too, see Smith 1999). The problem with this is that it creates an inescapable paradox: Papuans are framed as either unable to represent themselves because they don't have the skills to do so or as unable to represent themselves because having obtained those skills they no longer 'count' as 'authentic' indigenous people. Clearly, when a paradigm leaves no space for people to have the dignity of self-representation the paradigm needs to change.

Fourthly, Bishop's (1995, 1998, 2005) model lists legitimation. Schools are in the business of authorising particular pieces of knowledge as legitimate and others as not, through assessments as well as through correction and affirmation of particular responses in the classroom. What gives these processes authority? As discussed earlier, the ways in which people assess, evaluate and legitimate knowledge vary between different epistemological traditions (Tran this volume; Sillitoe 2010). For example, Sillitoe (2010) describes the linguistic markers that relate to the veracity of knowledge which are used by the Wola, a group of New Guinea highlanders, which include markers for whether a speaker or hearer of a given statement witnessed what is being discussed, and how far in the past it was, with increased trust being put in things which were witnessed by both speaker and hearer recently. Knowledge is legitimated through trust which is built on shared experience, rather than on trust in experts or authorities; there *is* no authority to adjudicate in disagreements regarding a particular piece of knowledge. These kinds of differences in the legitimation of knowledge are crucial to consider for schools, in which, traditionally, a teacher is presented as an expert (whose teaching draws on other experts) and children are expected to consider knowledge legitimate because it is given in that context, even if neither teacher nor child has had experience of whatever is being taught.

Fifthly, there is the question of accountability, which refers both to the question of who retains the power to assess and determine acceptable outcomes of schooling, and to the means by which schools are held accountable. One of the questions that foreign agencies investing in formal education need to ask is with whom does authority, and accompanying responsibility, for the education of Papuan children rest, and if it has been delegated in some measure to those working in schools, who delegated it? One answer, perhaps the most obvious, is that it is the parents of school pupils to whom schools should be accountable. Other potential contenders are the state, the foreign organisation associated with the school, the schools' funders, and the local sponsors of visas for foreign personnel. With so many potential pitfalls to introducing schools in the Papuan highlands, it is important to resolve to whom exactly a school is accountable and whether or not the relations of power that are implicit in that relationship of accountability are justifiable to the people the school exists to benefit.

A second aspect of the question of accountability is what mechanisms are used to hold the school accountable, and whether they make sense and are accessible to the relevant stakeholders. One criticism of the current state education system in Papua is that there are no practical mechanisms for monitoring and accountability. An incident that happened in a Walak speaking area of the Papuan highlands demonstrates the problem. In this area, there is an *SD* school and an *SMP* school just over half a kilometre apart from each other. In March 2013, the father of a young man who had attended both schools nailed up the doors and windows of the schools in broad daylight, triggering a meeting of the head teachers, teachers and parents to resolve the problems he was angry about. As I asked school children and other parents, who condoned this man's actions, what his motivations were, they explained that it is a typical practice at the *SD* school to change children's names at school, recording their pass results and school certificates in the wrong names. This creates chaos when children then try to register for *SMA* in Wamena. Some children's certificates are in other people's names, and some have even had their *aluak* changed, a name which indicates who their ancestors are, who they are obligated to and protected by, and who they can marry. Others report their ages being wrongly recording, leading to them being denied entry into *SMA* for being too young. There are even reports of males being recorded as females and vice versa. The reasons for this practice are not known. One of my Walak friends suggested that it is because the head teacher has to suddenly send off a list of results and does not want to take the time and effort to make sure all the names and corresponding results are correct. This does not, however, explain why some children report being consistently called by the wrong name in class by teachers.

Parents were also angry at the head teacher of the *SMP* school, who was apparently not providing *Surat Keterangan Hasil Ujian* (Letter Declaring Exam Results), a letter reporting that a student has passed her or his exams even though the certificates are still being processed, in time for students to enrol in *SMA*. The resolution meeting went ahead, but reports as to how successful this method of accountability was vary. I have been told that the head teacher of the *SMP* did not turn up to the meeting but that the necessary letters were consequently processed. Others have said that many students are still waiting for their *Surat Keterangan Hasil Ujian*. Apparently there has been no change in the practice of wrongly recording names and ages at the *SD* and, as far as parents know, the certificates with wrong names cannot now be changed. This example demonstrates the gaps between official methods of accountability, which parents either do not know about or do not want or feel able to use, and local methods of accountability, which achieve varying levels of success through unorthodox means. The head teachers are officially accountable to people who apparently do not know, and do not seem to have any means of knowing, about the problems parents and children are facing in these schools. Although this example is from a state school, the principle also applies to private schools. In any school where stakeholders are at such considerable linguistic, geographic and cultural distance from each other the questions of how a school is to be held accountable, and by whom, are crucial.

Paradoxes and Alternatives

This application of Bishop's (1995, 1998, 2005) model to analyse issues of power according to initiation, benefits, representation, legitimation and accountability in Papuan schools shows that the criticisms that indigenous (and especially Maori) scholars have made about both schools and academic research in other contexts are relevant to the Papuan context too. The first step to preventing the same mistakes being made in Papua is to recognise that schooling, like research, operates within relations of power, which have historically been managed in unjust ways. Non-Papuan people who do recognise that power imbalance may be inclined to adapt models of schooling that are familiar to them (to be more relevant and accessible to people living in the Papuan highlands) seeking, perhaps, to empower local people by encouraging them to get involved in the various processes of initiating and running a school. Although this may go some way towards addressing these issues, Bishop's quarrel with 'empowering' models of research (Bishop 1999; Smith 1999) is applicable to schooling too. If outsiders still control the initiation, representation, legitimation and accountability, as well as the extent to which local people can give their input and perspective, they perpetuate unequal power relations. Even approaches which are presented as collaborative by foreign agencies may fall into this trap of retaining control over the terms of collaboration, however well-intentioned their efforts.

An alternative is for the control and power over the schools and their processes to be exclusively in the hands of indigenous Papuan highlanders. In theory, if a particular community wanted access to the knowledge and skills normally taught in schools, it could articulate its goals and design, or commission the design of, a formal education institution which met its needs, and determine how it would operate and who would be involved. This would not exclude the input of non-indigenous people. The difference is that the input would be on the highlanders' terms and the non-indigenous people would be accountable to the community they were working for. In reality, however, the political and historical context is such in Papua that to initiate and run such a school necessitates the ability to negotiate bureaucratic systems and funding streams, as well as to access and communicate with people who would be able (and willing) to teach these new skills on terms set by the highlanders, which in itself would require having the knowledge and skills to operate on foreign terms.

This paradox highlights a central dilemma in providing education for indigenous Papuans. It is crucially important to argue for the value and legitimacy of indigenous knowledge and to recognise that people can be educated without being schooled. But doing so does not change the present reality that lack of access to formal education leaves indigenous Papuans on the losing end of unequal power relations. Formal education is a system in its own right, one which, among other things, depends heavily on the written word, creates and relies upon experts, strongly values the accumulation of knowledge over generations and legitimates knowledge produced through the scientific method. The system, and the knowledge it produces, are powerful. As Bernard (2011: 12) asserts, 'scientifically produced knowledge is effective – it lets us control nature, whether we're talking about the weather, or disease, or our own fears, or buying habits'. The dominance of formal education as a means to producing knowledge is so widespread that access to schooling is considered a human right and 'universal primary education' is a global development goal (United Nations 2013; UNESCO 2012).

This dominance leaves indigenous Papuans too frequently forced to choose between unequal power relations due to lack of schooling and unequal power relations within schooling, with potentially disastrous consequences (see Nichol this volume). The situation is analogous to that of minority language speakers who know that to bring their children up speaking only the dominant language is to alienate them from their heritage and identity as well as to contribute to the death of a language and the knowledge and meaning embedded in it, but to bring them up speaking only their mother tongue is to severely disadvantage them in pursuing social, political and economic engagement with the wider community (Bishop and Glynn 1999). Consequently, many children grow up bilingual, using different languages in different domains and able to switch fluently between them according to context. Bilinguals rarely keep their languages completely separate, but are able to draw from one to enrich the other, translate between them, and codeswitch between them to achieve a variety of communicative purposes. Bilinguals are usually bicultural too, and are competent in behaving and communicating in culturally appropriate ways within each language (Baker 2011). Is it conceivable, then, that children could become bi-educated, just as so many have learned to be bilingual? There has been a significant amount of research done on bicultural education, usually embedded into research on bilingual education that recognises that language and culture are related and seeks to ensure space for the cultures associated with both languages in the formal bilingual education system. That body of research does not address the question of whether children could successfully learn within with two different, separate education systems simultaneously, learning to use different knowledge systems in different domains, to switch fluently between them according to context, and ultimately, to be able to draw from each to enrich the other.

It is likely that many Papuans have learned to be fluent within two separate knowledge systems, negotiating the conflicts between them with very little support and perhaps even at great personal cost. The development of bi-educationalism, by which I mean formal education and indigenous education intentionally accommodating and supporting each other, would necessitate partnerships between Papuans and people from foreign agencies around a shared purpose in which the strengths that each contributor, whether indigenous or foreign, has to offer are recognised as invaluable to the fulfilment of the partnership's purpose. This could be a potential way through the impasse of unequal power relations in initiation, benefits, representation, legitimation and accountability in education. For example, if the purpose is to establish a school to serve the children of a particular highland community, which educates pupils in Western knowledge without threatening their continued education in indigenous knowledge, non-locals may bring knowledge about school models from other contexts, knowledge of some of the new skills that may be desirable to the indigenous community (such as reading, writing and speaking English), an understanding of foreign epistemologies and access to funding, among other things. Indigenous members of the partnership may bring knowledge of models of education from their own context and of the skills and knowledge which their children need to thrive locally, as well as an understanding of local epistemologies and access to local resources, among other things. Both partners would need to be committed to supporting different forms of education, learning from each other, and negotiating inevitable conflicts between the two education systems.

It is essential that the outcomes of such a partnership genuinely benefit Papuan highlanders. Given a commitment to mutual respect, genuine mechanisms for accountability, and honest and transparent communication, the negotiation of wider benefits such as the publication of knowledge for a wider audience, or the formation of parallel partnerships in other contexts in which Papuans are the visitors, does not seem to me to be mutually exclusive with respect for the power of an indigenous community. In a true partnership, both hosts and visitors will benefit, and there may even be wider benefits such as new and creative approaches to global problems. The challenge is to achieve mutuality, which is no easy feat given the historical context and the different partners' dissimilar traditions of leadership, knowledge and success. Foreign partners must be accountable to, and remain within the boundaries set by, local partners. However, foreign partners will not, and should not, participate in ways that violate their own values or rights either. If shared values and goals can be set, and accessible mechanisms for accountability that make sense to both partners can be agreed upon, I think there is potential for exciting collaborative work.

This approach is similar to the Maori research metaphor of a *whanau* of interest in which 'the enormously important task of recognizing the relative *tapu* (specialness; being with potentiality for power) and *mana* (power) of the two sides, the hosts and the visitors', occurs during the *hui* formal meeting (Bishop 1998: 206). One part of the *hui* is the offering, by the visitors, of a gift towards the cost of the meeting, which they present to be considered, and accepted or rejected by the hosts. This process manifests the hosts' right to reject the visitors' participation, and is a powerful image of their right to self-determination. If the visitors' offering is accepted, the whole *whanua*, visitors and hosts, own the agenda and problems for consideration in the *hui* and both share the cost and responsibility of addressing them. Although the model of a *whanau* is specific to the Maori and should no more be imported wholesale to the Papuan context than any other foreign model, it does provide an example of a partnership process that is fashioned on an indigenous concept, in which neither indigenous nor non-indigenous participants are required to give up power to work together (for further examples see Nichol this volume; Borden 2013; Lipka et al. 2013; Bishop and Glynn 1999). Partnerships formed on such principles, but drawing on Papuan metaphors, could provide a way forwards.

There are many difficulties associated with the kinds of partnerships I'm proposing. For instance, if partnerships for education are to draw on indigenous concepts and accommodate local forms of knowledge and learning, they will need to be unique to each context. The issues discussed in this paper have focused mostly on individual schools, but most foreign investment in Papuan education is on a national scale, through the Indonesian and provincial governments. Private individual schools only reach a limited number of children, and the resources needed to provide schooling adapted to each indigenous context in Papua are enormous. However, private schools can be influential as models and partners for change in the state system (Government of Papua Province et al. 2009; USAID 2009; UNDP 2005). Even on a school-by-school basis finding ways to work which are acceptable to all partners is difficult. Education is inherently connected to beliefs and there are many Papuans who would argue that indigenous knowledge and education can't be separated from indigenous spiritual beliefs, beliefs which are at odds with most mission agendas and many secular schooling agendas. Negotiating the wider political issues, and dealing with the challenges and paradoxes drawn out in this chapter are also among the many potential obstacles to building a successful team with shared goals.

Nonetheless, an alternative to what has been done in the name of education in other parts of the world is needed in Papua. In this chapter, I have outlined the history of schooling in Papua and the problems of the current formal education system. I have argued that foreign agencies wanting to help solve those problems must recognise the issues of power involved in formal education, which I have analysed, after Bishop (1995, 1998, 2005), according to initiation, benefits, representation, legitimation and accountability. I have then examined potential alternatives to traditional models in which outsiders control each of these areas, and argued for collaborative partnerships. I disagree with those who suggest that only indigenous people are equipped to solve the dilemmas discussed. I think both anthropologists and foreign educators who recognize the ways that research and education have historically operated hierarchically, and who are committed to working collaboratively, can have a role to play in developing more just forms of education for indigenous Papuans.

References

Agrawal, A. 1995. Dismantling the Divide Between Indigenous and Scientific Knowledge. *Development and Change*, 26(3), 413–39.

Ajamiseba, D.C. 1987. Primary Education in Irian Jaya: A Qualitative Background Report with a Proposed Strategy. *Irian, Bulletin of Irian Jaya*, 15, 3–17.

Alexander, R.J. 2001. Border Crossings: Towards a Comparative Pedagogy. *Comparative Education*, 37(4), 507–23.

Ascher, M. and Ascher, R. 1986. Ethnomathematics. *History of Science*, 24(2), 125–44.

Badan Pusat Statistik Republic of Indonesia 2013. *Percentage of Population who are Illiterate by Province and Age Group, 2003–2012* [Online]. Jakarta: Badan Pusat Statistik. Available at: http://www.bps.go.id/eng/menuTablephp?tabel=1&kat=1&id_subyek=28 [Accessed 21 June 2013].

Baker, C. 2011. *Foundations of Bilingual Education and Bilingualism*. Bristol: Multilingual Matters.

Bernard, H.R. 2011. *Research Methods in Anthropology: Qualitative and Quantitative Approaches.* Plymouth: AltaMira.

Bishop, R. 1995. *Collaborative Research Stories: Whakawhanaungatanga.* PhD, University of Otago. Available at: http://otago.ourarchive.ac.nz/handle/10523/531, accessed: 21 June 2013.

—— 1998. Freeing Ourselves from Neo-Colonial Domination in Research: A Maori Approach to Creating Knowledge. *International Journal of Qualitative Studies in Education*, 11(2), 199–219.

—— 1999. Kaupapa Maori Research: An Indigenous Approach to Creating Knowledge, in *Maori and Psychology: Research and Practice*, the proceedings of a symposium sponsored by the Maori and Psychology Research Unit, edited by N. Robertson. Hamilton: Maori and Psychology Research Unit, University of Waikato.

—— 2003. Changing Power Relations in Education: Kaupapa Māori Messages For 'Mainstream' Education in Aotearoa/New Zealand. *Comparative Education*, 39(2), 221–38.

—— 2005. Freeing Ourselves from Neo-Colonial Domination in Research: A Kaupapa Maori Approach to Creating Knowledge, in *The Sage Handbook of Qualitative Research*, edited by N.K. Denzin and Y.S. Lincoln. Thousand Oaks: Sage Publications Inc., 109–38.

Bishop, R. and Glynn, T. 1999. *Culture Counts: Changing Power Relations in Education.* Palmerston North: The Dunmore Press Ltd.

Borden, L.L. 2013. What's the Word for … ? Is There a Word for … ? How Understanding Mi'kmaw Language can Help Support Mi'kmaw Learners in Mathematics. *Mathematics Education Research Journal*, 25, 5–22.

Butt, L. 1998. *The Social and Political Life of Infants Among the Baliem Valley Dani, Irian Jaya.* PhD, McGill University. Available at: http://www.papuaweb.org/dlib/s123/butt/phd.pdf, accessed: 15 June 2013.

Cram, F. 1992. *Ethics in Maori Research: Working Paper* [Online]. Auckland: University of Auckland. Available at: http://researchcommons.waikato. ac.nz/bitstream/handle/10289/3316/Cram%20-%20Ethics%20in%20Maori. pdf?sequence=1, accessed: 26 June 2012.

Coolangatta Statement on Indigenous Rights in Education 1999. The Coolangatta Statement on Indigenous Rights in Education. *Journal of American Indian Education*, 39(1), 52–64.

Draft Peraturan Daerah Provinsi Papua Concerning The System of Education in Papua Province 2002. Jayapura: Papua House of People's Representatives and the Governor of Papua Province. Available at: www.papuaweb.org/dlib/lap/ sullivan/perdasi-perdasus/10.rtf, accessed: 16 June 2013.

Duijnstee, A.J.H. 1972. Inland Village Education: A Suggestion for Change. *Bulletin of West Irian Development*, 1(2), 22–7.

Frederick, W.H. and Worden, R.L. (eds) 2011. *Indonesia: A Country Study*. Washington: Library of Congress.

Government of Papua Province, World Bank and United Nations in Indonesia 2009. Papua Accelerated Development Needs Assessment (PADNA), Vol. 1: Report on Recommendations and Action Plan Interventions, November 2009. Available at: http://www.un.or.id/documents_upload/publication/PADNA%20 VOL1%20final%20draft-091106.pdf, accessed: 15 June 2013.

Harvey, G. 2003. Guesthood as Ethical Decolonising Research Method. *Numen*, 50, 125–46.

Hermes, M. 2005. 'Ma'iingan is Just a Misspelling of the Word Wolf': A Case for Teaching Culture Through Language. *Anthropology & Education Quarterly*, 36(1), 43–56.

Heshusius, L. 1994. Freeing Ourselves From Objectivity: Managing Subjectivity or Turning Toward a Participatory Mode of Consciousness? *Educational Researcher*, 23(3), 15–22.

Ismail, S.M. and Cazden, C.B. 2005. Struggles for Indigenous Education and Self-Determination: Culture, Context, and Collaboration. *Anthropology & Education Quarterly*, 36(1), 88–92.

Jorgensen, R. and Wagner, D. 2013. Mathematics Education With/For Indigenous Peoples. *Mathematics Education Research Journal*, 25, 1–3.

Kaomea, J. 2005. Indigenous Studies in the Elementary Curriculum: A Cautionary Hawaiian Example. *Anthropology & Education Quarterly*, 36(1), 24–42.

Lipka, J., Wong, M. and Andrew-Ihrke, D. 2013. Alaska Native Indigenous Knowledge: Opportunities for Learning Mathematics. *Mathematics Education Research Journal*, 25, 129–50.

Manuelito, K. 2005. The Role of Education in American Indian Self-Determination: Lessons from the Ramah Navajo Community School. *Anthropology & Education Quarterly*, 36(1), 73–87.

May, S. and Aikman, S. 2003. Indigenous Education: Addressing Current Issues and Developments. *Comparative Education*, 39(2), 139–45.

Mollet, J.A. 2007. Educational Investment in Conflict Areas of Indonesia: The Case of West Papua Province. *International Education Journal*, 8(2), 155–66.

Munro, J. 2009. Dreams Made Small: Humiliation and Education in a Dani Modernity. PhD, The Australian National University. Available at: http://www.papuaweb.org/dlib/s123/munro/_phd_no_images.pdf, accessed: 12 June 2013.

—— 2013. The Violence of Inflated Possibilities: Education, Transformation, and Diminishment in Wamena, Papua. *Indonesia*, 95, 25–46.

Nakata, M. 1998. Anthropological Texts and Indigenous Standpoints. *Australian Aboriginal Studies*, 2, 3–12.

Nichol, R. 2011. *Growing up Indigenous: Developing Effective Pedagogy for Education and Development.* Rotterdam: Sense Publishers.

Oikonomos Foundation 2012. *Koinonia, Wamena Nursery and Primary Education* [Online]. Oikonomos Foundation. Available at: http://www.oikonomos.org/Page/sp323/ml2/from_sp_id=317/nctrue/system_id=12724/so_id=1490/Index.html, accessed: 5 July 2012.

Porsanger, J. 1994. An Essay about Indigenous Methodology. *Nordlit*, 15, 105–20.

Sarangapani, P.M. 2003. Indigenising Curriculum: Questions Posed by Baiga Vidya. *Comparative Education*, 39(2), 199–209.

Scribner, S. 1977. Modes of Thinking and Ways of Speaking: Culture and Logic Reconsidered, in *Thinking – Readings in Cognitive Science*, edited by P.N. Johnson-Laird and P.C. Watson. Bristol: Cambridge University Press, 483–500.

Sekolah Papua Harapan 2012. *Sekolah Papua Harapan* [Online]. Sentani. Available at: www.papuaharapan.org, accessed: 25 June 2012.

Serasi 2012. *Pivotal Point for Education in Papua* [Online]. Serasi, via International Relief and Development, for USAID. Available at: http://www.serasi-ird.org/index.php/activities/79-pivotal-point-for-education-in-papua, accessed: 21 June 2013.

Sillitoe, P. 2010. Trust in Development: Some Implications of Knowing in Indigenous Knowledge. *Journal of the Royal Anthropological Institute*, 16(1), 12–30.

Smith, L.T. 1999. *Decolonizing Methodologies: Research and Indigenous Peoples.* London: Zed: University of Otago Press.

—— 2005. Building a Research Agenda for Indigenous Epistemologies and Education. *Anthropology & Education Quarterly*, 36(1), 93–5.

Somba, N.D. 2009. Papuan Children Show Off their Math Skills [Online]. Jakarta: *The Jakarta Post*. Available at: http://www.thejakartapost.com/news/2009/09/16/papuan-children-showoff-their-math-skills.html, accessed: 26 June 2012.

Special Autonomy for the Papua Province, Bill of Law of the Republic of Indonesia No. 21 2001. Jakarta: House of People's Representatives of the Republic of Indonesia. Available at: http://www.papuaweb.org/goi/otsus/files/otsus-en.html, accessed: 21 June 2013.

Spicer, P. 1972. Some Thoughts on Educational Aims in the Highlands of Irian Barat. *Bulletin of West Irian Development*, I(2), 16–21.

Sulthani, L. 2012. *Remote Indonesian School Fights Illiteracy with Bows* [Online]. Available at: http://www.trust.org/alertnet/news/remote-indonesian-school-fights-illiteracy-with-bows, accessed: 15 June 2013.

Surya Institute 2009. *Surya Institute Vision and Mission* [Online]. Jakarta: Surya Institute. Available at: http://www.suryainstitute.org/en/content/view/12/30/, accessed: 21 June 2013.

Te Awakotuku 1991. *He Tikanga Whakaaro: Research Ethics in the Maori Community.* Wellington: Manatu Maori.

Timmer, J. 2005. Decentralisation and Elite Politics in Papua. *State, Society and Governance in Melanesia*, Discussion Paper 2005/6, ANU, 1–18.

UNDP 2005. *Papua Needs Assessment: An Overview of Findings and Implications for the Programming of Development Assistance.* Available at: http://www.undp.or.id/papua/docs/PNA_en.pdf, accessed: 20 August 2005.

—— 2012. *Fighting Illiteracy in Indonesia's Papua Region* [Online]. UNDP in Indonesia. Available at: http://www.undp.org/content/indonesia/en/home/ourwork/povertyreduction/successstories/fighting-illiteracy-in-indonesias-papua-region/, accessed: 15 June 2013.

UNESCO 2012. *Education for All Goals* [Online]. UNESCO. Available at: http://www.unesco.org/new/en/education/themes/leading-the-international-agenda/education-for-all/efa-goals/, accessed: 21 June 2013.

UNICEF Indonesia 2009. *An Interview with Mr James Modouw, Head of the Provincial Education Office* [Online]. UNICEF. Available at: http://www.unicef.org/indonesia/reallives_11152.html, accessed: 21 June 2013.

United Nations 2013. *We Can End Poverty. 2015 Millennium Development Goals* [Online]. New York: United Nations. Available at: http://www.un.org/millenniumgoals/, accessed: 21 June 2013.

USAID 2009. *Papua Assessment, USAID/Indonesia, November 2008–January 2009.* Available at: http://indonesia.usaid.gov/documents/document/document/351, accessed: 13 September 2010.

—— 2011. *Papua Primary School Program Transforms Community* [Online]. Jakarta: USAID Indonesia. Available at: http://indonesia.usaid.gov/en/USAID/Article/608/Papua_Primary_School_Program_Transforms_Community [Accessed 21 June 2013].

Varenne, H. 2007. Difficult Collective Deliberations: Anthropological Notes Toward a Theory of Education. *Teachers College Record*, 109(7), 1559–88.

World Bank 2009. *Investing in Indonesia's Education at the District Level: An Analysis of Regional Public Expenditure and Financial Management.* Available at: http://ddp-ext.worldbank.org/EdStats/IDNper09.pdf, accessed: 21 June 2013.

Index